are globally facing an existential climate crisis and yet governments, compa-
and citizens are struggling to move fast enough to avoid the worst impacts
lobal warming. Bringing scholarly research perspectives to understanding
physical and transitional risks that climate change poses for business will be
tial to making progress. Dowbiggin's new book is a welcome intervention
is conversation."

—Sarah Kaplan, *Distinguished Professor, Rotman School of Management,*
University of Toronto, Canada

Climate Risk and Busines

"W
nies
of
the
esse
in t

Anna Dowbiggin

Climate Risk and Business

New Challenges for Organizations

Anna Dowbiggin
University of Guelph-Humber
Toronto, ON, Canada

ISBN 978-3-030-78243-6   ISBN 978-3-030-78244-3   (eBook)
https://doi.org/10.1007/978-3-030-78244-3

This Palgrave Macmillan imprint is published by the registered company Springer Nature Switzerland AG
The registered company address is: Gewerbestrasse 11, 6330 Cham, Switzerland

PREFACE

The purpose of this book is to conceptualize the transition challenges awaiting business as a result of climate risk. This is done through my tentative fivefold proposition in this book that organizations will rearrange organizational priorities and management practices to reduce risk exposures created by climate risk.

I regard this holistic academic treatment of climate risk as urgent and critical, given a fragmented literature base, low empirics on decarbonization practices, and the disappointing reported business response to reduce carbon across value chains.

Urgent, because we are running out of time to reduce global atmospheric GhG concentration levels. Critical, because an 'all-in' view of business mitigation effort has an enormous role to play in course-correcting the climate trajectory we face.

This book is written for academic audiences primarily. My attention is given to theory as well as practice, to paint a view of empirics that may or may not be supported by theory, and to demonstrate where scholarly enlivenment opportunities are located. Readers may note some extant theoretical speculations may need refurbishments to account for current business conditions. I view this as a source of academic contribution needed in both literature and in the teaching of responsible business in business school curricula.

Furthermore, this book topic is written with a dominant risk management lens to reflect what I call the emergent 'ontological standardization' of climate risk. Definite ideas about what climate risk is and its prescriptions (generally speaking) are promulgated by the Task Force on Climate-Related Risk and the Financial Stability Board.

This is a significant departure from prior management theory on organizational climate response and is reminiscent of conversations I have had with practitioners. In those instances, I had the privilege of lively and protracted business conversations about the difference between climate change and climate risk. Those differences, I hope, emerge as a secondary outcome of reading this book—to move past prior climate narratives, and to reconceptualize the climate emergency as a mainstream and multidimensional strategic business risk of ultimate importance to everyone.

Toronto, Canada   Anna Dowbiggin

Acknowledgments

I wish to thank my colleagues, friends, and family who contributed to the development of this book. I am grateful to Prof. Devi Jankowicz at Edinburgh Business School where I began to think about climate risk as a research agenda in 2017. The insights of colleagues at the US Academy of Management, Chatham House in the UK and the many business practitioners who are working on carbon strategies within the industry, are greatly acknowledged for their support. I am grateful for your insights, wisdom—and friendships. I am also grateful for the inspiration routinely provided to me by the students I teach and mentor in business research and climate studies. I am also indebted to Ph.D. candidate William Little who assisted in the production of the manuscript, and to the editors at Palgrave Macmillan who took me on as the author for this important book topic.

Contents

About the Author

Dr. **Anna Dowbiggin** is a Business Professor with University of Guelph Humber and has taught at Ryerson University, Humber Institute of Technology and Advanced Learning, and York University in Canada.

Her research interests are focused on climate risk management, energy transition, and decarbonization in industry. She currently directs a Climate Risk Lab initiative with 10 companies undergoing transformational carbon reduction processes. She is an invited academic delegate to COP 26 in Glasgow, 2021.

She earned her Doctorate in Business Administration at the Edinburgh Business School at Heriot-Watt University in Scotland. Her dissertation research was on risk management cognition among chief executive officers in the electricity sector in Canada. Her dissertation earned her several academic nominations and public scholarship mentions in the business press in the UK and Canada.

She further benefits from postdoctoral work in risk management at McMaster University and sits on numerous boards as an advisor and climate risk expert. She also benefits from a long career as a business executive with US corporations in operations and general management.

Abbreviations and Acronyms

3LoD	3 Lines of Defense Model
ACT	Assessing Carbon Transition
BPA	Bisphenol A
CAS	Casualty Actuarial Society
CDP	Carbon Disclosure Project
CDSB	Climate Disclosure Standards Board
COSO	Committee of Sponsoring Organizations of the Treadway Commission
DC	Dynamic Capabilities Perspective
DFS	Department of Financial Services
ERM	Enterprise Risk Management
ESG	Environmental, Social, and Governance
FRB	Future Time Reference
FSB	Financial Stability Board
FTSE	Financial Times Stock Exchange
GDP	Global Domestic Product
GhG	Green House Gas
GRI	Global Reporting Initiative
ICRMS	Internal Control and Risk Management Systems
IPCC	Intergovernmental Panel on Climate Change
KRF	Key Risk Factor
LCA	Life Cycle Assessment
LGIM	Legal and General Investment Management
MSCI	Morgan Stanley Capital International
NRBV	Natural Resources-Based View
RBV	Resource-Based View

RIMS	Risk and Insurance Management Society
SASB	Sustainability Accounting Standards Board
SBTi	Science Based Targets Initiative
SDG	Sustainable Development Goals
SEC	Security and Exchange Commission
TCFD	Task Force on Climate-Related Financial Disclosures
UNPRI	United Nations Principles for Responsible Investment
UNSDG	United Nations Sustainable Development Goals
VRIN	Valuable, Rare, Imperfectly Imitable, Non-substitutable

LIST OF FIGURES

CHAPTER 1

Climate Risk and Organizational Challenges

Abstract The introduction of new risk categories to an organization represents a major transition in organizational priorities and management practices. The challenges of climate risk to business will most likely be varied, vast, and complex. A tentative fivefold proposition of how climate risk exposures will challenge business is introduced in this chapter.

Keywords Climate risk · Mitigation · Business transition · Decarbonization · Low carbon initiatives · Net zero

1.1 INTRODUCTION

Climate change has exposed business to a complex set of climate risks that must now be addressed through bold mitigation action (Carney, 2021; TCFD, 2017, 2020).

The growing evidence of severe climate change impacts on human life and the global economy has created the urgent need for a mid-century strategy of climate mitigation to prevent future climate destabilization. In literature, mitigation has been summarized as biophysical and/or political economic response by organizations (Wittneben et al., 2012). Climate risk management expectations by financial and business governance groups, suggest multiple and varied deployments of mitigation

A. Dowbiggin, *Climate Risk and Business*,
https://doi.org/10.1007/978-3-030-78244-3_1

1

strategies are necessary to combat climate-related risks (Flusberg et al., 2017). This underscores the need to recognize that organizational climate risk challenges are holistic, complicated, and defy prior business assumptions about managing risks. Furthermore, the goal of global mitigation action is to achieve a net zero target by 2050, the notional threshold of atmospheric GhG emissions to stabilize climate sensitivity to further GhG levels. How firms expect to initiate and accelerate their own organizationally bounded mitigative approaches to reach net zero targets, will transform cognitions and risk beliefs, organizational resources, informal and formal risk management systems, and the way in which firms conceptualize and disclose future dependencies on climate.

1.2   Risk Exposures

The effects of climate change have been already visible as catastrophic weather events but are now beginning to impact assets which have a market price, making the scale of climate risk more tangible. As Mark Carney, former chair of the Central Bank of England (CBE), comments, 'Climate change is setting in train a vicious cycle in which rising sea levels and more extreme weather are damaging property, forcing migration, impairing assets and reducing the productivity of work' (Carney, 2021, p. 263).

Global climate policy groups, i.e., the Intergovernmental Panel on Climate Change (IPCC) and financial groups concerned with the stability of the global economy, i.e., the Financial Stability Board (FSB), assert that current atmospheric greenhouse gas (GhG) concentration levels must be urgently and radically reduced by every and all global constituents. Global constituents include market actors in all sectors of the economy which produce GhG from their own production processes, the energy systems they rely on, and from the products and services they produce. The biggest industrial contributors to GhG emission production are organizations with emission-generating industrial processes, built environments which consume energy and heat, power producer groups which supply energy, and food and agriculture industries which also produce emissions. Investment parties which invest in carbon producing companies, e.g., private equity, asset management firms, as well as insurance and lending institutions, are exposed to climate risk through portfolio engagement.

Small- and mid-sized companies are not exempt from climate mitigation scrutiny, as many have carbon footprints of their own, and are suppliers to larger industrial groups with sizeable carbon footprints of their own. Inaction on the part of any of these business groups to recognize, assess, and initiate climate risk mitigative strategies, will create massive loss exposures throughout business value chain and the global economy.

1.3   TRANSITION THEORY

The likelihood of massive loss exposure in business sectors stems from climate-related risk, i.e., climate risk, given its implication for business organizations and the financial groups which invest in them. The strategic and financial implications of climate risk exposure pose major challenges for business because they are fundamentally different from other business risks. In this book, I posit that the transition companies will undertake to secure alignment with evolving and stringent carbon reduction expectations and formal climate risk regime requirements, will impact business through multiple transition challenges. Underscoring these challenges is the recognition of risk regimes and business transition, consistent with Geels' (2002) socio-technical regime rules in transition theory. As Geels asserts, regimes refer to the dominant rules that preserve the stability of existing systems, including 'shared beliefs,' 'capabilities,' and 'user practices'; transitions refer to the shift from one given regime to another, that requires modifications in these three different dimensions (Geels, 2002). In this regard, I tentatively propose that climate risk exposures are forcing a global business transition that is disruptive of dominant business as usual (BAU) logics that normally preserve the stability of organizational life.

1.4   RISK RESPONSE OPTIONS

According to the American Meteorological Society (AMS), risk response options to reduce climate risk consist of three proactive strategies: (1) mitigation—the efforts to reduce firm-level GhG emissions; (2) adaptation—the efforts to increase organizational capacity to cope with changes in climate; and (3) geoengineering—the efforts to deliberately manipulate the earth system that is intended to counteract some of the impacts of GhG emissions (Stenhouse et al., 2014). Geoengineering and its performance parameters are underdeveloped and may not represent a reasonable option for business at this position in time; however, we can accept that general organizational responses fall into either (or both) mitigation and adaptation response categories. Given the approximation of atmospheric GhG concentration and the probable trajectory of dangerous global warming with further increases in GhG concentration, mitigation through the wholesale decarbonization of industry appears to be the only immediate and practical solution for business.

Understandings of climate mitigation expressed in 2014 were defined by the Intergovernmental Panel on Climate Change (IPCC) as:

> Mitigation—reducing climate change—involves reducing the flow of heat-trapping greenhouse gases into the atmosphere, either by reducing sources of these gases (for example, the burning of fossil fuels for electricity, heat or transport) or enhancing the "sinks" that accumulate and store these gases (such as the oceans, forests and soil). The goal of mitigation is to avoid significant human interference with the climate system and stabilize greenhouse gas levels in a timeframe sufficient to allow ecosystems to adapt naturally to climate change, ensure that food production is not threatened and to enable economic development to proceed in a sustainable manner. (IPCC, 2014, p. 4)

Seven years later the IPCC stressed the need for mitigation by business:

> Climate change mitigation is achieved by limiting or preventing greenhouse gas emissions and by enhancing activities that remove these gases from the atmosphere. Greenhouse gases can come from a range of sources and climate mitigation can be applied across all sectors and activities. These include energy, transport, buildings, industry, waste management, agriculture, forestry, and other forms of land management. (IPCC, 2021, p. 2)

1.5   Mitigation Approaches

Mitigation, in practitioner language as it applies in corporate risk practice, is a risk response of prevention and/or loss reduction. Firms aspiring to the stringent requirement of a net zero target of global emissions, must find options to reduce GhG in both demand and supply sides of business operations, covering all three scope zones in the business model. Unlike climate adaptation response which provides localized and immediate term benefits to business, climate mitigation is concerned with (often costly) carbon reduction activities today which provide benefits in the longer term. The problem with adaptation as an exclusive risk response, is that it presupposes that adapting to climate change is a sufficiently appropriate solution. Adaptation is adapting to what is already happening and learning to live with it. It does not reduce, prevent, or mitigate against the source of the problem in the first instance. As Nordhaus (2013) and many scholars comment, 'We cannot adapt our way out of climate change.'

In practice, examples of mitigation approaches by industry include: emissions reduction (Bui & de Villiers, 2017) new low carbon products and operational processes (Cadez & Czerny, 2016; Kolk & Pinkse 2008) low carbon logistics (de Sousa Jabbour et al., 2019) supplier involvements in decarbonization (Lee et al., 2012). While the above examples are not exhaustive, they provide a preliminary view of applications in intra-organizational systems. The link between mitigation effort and the global carbon budget, is discussed next.

1.6 Global Carbon Budget

Rapid industrialization and global growth have increased the level of GhG, especially CO_2 in the atmosphere at an alarming rate. It took 250 years to burn the first half trillion tons of carbon and based on current trends, the next half trillion will be released into the atmosphere in less than 40 years (Allen et al., 2009). But it is not emissions per se that are the source of concern; it is the concentration of CO_2 and other GhGs in the atmosphere that have risen to levels exceeding 412 ppm, creating an 'enhanced greenhouse effect' and an intractable rise in mean temperatures on the earth's surface. Increasing the atmospheric concentration of CO_2 by what seems a tiny fraction (from 280 to 560 ppm, as one example) is projected to increase average surface temperatures of the planet by around 3 °C, or 5.5. °F (Nordhaus, 2013).

Basic intuition behind climate science tells us that the concentration of atmospheric GhG is closely related to the pace of global warming. From this proposition, scientific calculations made by the Intergovernmental Panel on Climate Change (IPCC) and its scientific research consortium of 650 partnerships around the world have been able to estimate a 'global carbon budget'—that is, how much CO_2 can still be released into the atmosphere before earth systems exceed different temperature thresholds.

Carbon budget analysis suggests that limiting temperature increases of 1.5 °C from pre-industrial levels keeps the earth's climatic and natural systems from tipping into a dangerous, intractable, feedback loop. The point here is that limiting anthropogenic emissions to that scale, to the extent the balance between the amount of greenhouse gas produced and the amount removed from the atmosphere will produce a 'net zero' balance, or the achievement of 'carbon neutrality.'

1.7   STRINGENCY OF NET ZERO TARGET

The stringency of a net zero target means that new and ambitious options to reduce emissions will be required of business, across different sectors, in radically different formats from those used previously.

In the energy sector, for example, coal or natural gas-based public energy systems in India, China, Australia, and the United States (where coal and natural gas still represent a sizeable proportion of the energy generation mix) will likely pursue a range of demand side adjustments such as efficient energy demand provision, materials efficiency, circular economies, and reduced consumption of goods (International Energy Agency, 2019). On the supply side, options such as hydrogen or biogas blending or switching, nuclear commissioning and deep integration of renewable solar and wind energy power sources are options available to business.

In industrial sectors which emit 17 Gt of CO_2e annually, or 32% of the global total, GhG emissions reduction may be equally ambitious and complicated.

As just one example, heavy industries such as cement manufacturing, plastics, aluminum, and chemicals contribute sizable proportions of industrial emissions through their carbon-intensive manufacturing processes (Carney, 2021; Pye et al., 2021).

Despite these calculations, the global carbon budget is still being drawn down and business is not on track to meet temperature goals for 2050. Temperature goals were set in the 2015 Paris Agreement, which calls for the adoption of a mid-century strategy of radical GhG reduction to mitigate against further planetary loss (The Whitehouse, 2017). It implores private sector actors to radically reduce anthropogenic GhG emissions to self-determined progress milestones leading up to 2050, i.e., 2025 and 2040, as is the case for Canada and the European Union, at the time of writing.

Today, seven years after the Paris Agreement, and four years after the latest dire warning of the IPCC 5th Assessment Report (ARC), policy mobilization, mounting regulatory requirements, and investor sentiments are altering business environments (Nordhaus, 2013). Climate legislation, as-of-now voluntary reporting frameworks and increasing regulatory interest in leading countries such as Belgium, Canada, Chile, Costa Rica, France, Japan, Germany, New Zealand, the United Kingdom, and the United States, have amplified the global decarbonization imperative (TCFD, 2020).

For business this means that these external pressures for climate risk management, are driving the requirement of companies to address climate risk in its various manifestations, rapidly and radically to unprecedented timelines. Sector approaches to customizing climate risk assessment approaches including carbon accounting methodologies are underway. Civil society actors, i.e., Swedish youth activist Greta Thunberg, have activated new dialogues on the social cost of carbon, and stakeholder activists have instigated due diligence appraisals of carbon reduction claims made by business groups.

1.8   MASSIVE LOSS EXPOSURES

The problem for business, is that any failure to adequately transition to a net zero levels of carbon will expose many firms to potentially significant losses. Climate risk is likely to make many existing business models unviable through severe disruptions in supply chains and destruction of assets (TCFD, 2017). Discontinuities in business operations due to extreme 'acute,' i.e., flash floods or droughts, or the escalating 'chronic' manifestations of climate events, i.e., gradual mean temperature increases, are driving forces for potential new business approaches to reduce climate risks. Furthermore, business losses are expected to impact the full scope of the business value chain, classified by the Green House Gas Protocol in three conceptual scope zones of business activity as follows: (1) Scope 1—direct emissions from owned or controlled sources; (2) Scope 2—indirect emissions from the generation of purchased electricity, steam, heating, and cooling consumed by the reporting company; and (3) Scope 3—all other indirect emissions that occur in a company's value chain (Carbon Trust, 2021).

Inaction or insufficient decarbonization for example, of core operations in Scope 1, supply chains and purchased energy in Scope 2, and its products and services in Scope 3, exposes businesses to additional secondary and cascading systemic risks. Those risks relate to what climate economists call 'the disorderly economic transition' that produces potential regulatory, legal, market, and reputational loss exposures for companies. For example, corporate failure to initiate low carbon initiatives in alignment with the science-based targets (SBTi), of the Carbon Disclosure Project (CDP), will have negative implications for corporate reputation. Failure to accurately report climate risks in well-defined financial filing reports will likely trigger regulatory response and legal liability action. Legal liability risk exposure is created when firms are known to have full knowledge of their own climate risk exposures but do not attend to them; in practice according to climate legal experts, climate liability risk is as simple as 'not reporting on climate risk' (Zizzo, 2017). Market risk exposures and potential financial losses will arise when market preferences for decarbonized assets increase, relegating all else to the 'stranded asset' heap. Market risk of this kind stands to create a zero-sum game of winners and losers based on company carbon performance. In other words, should business fail to initiate, transition, and sustain a net zero carbon strategy over time, its future is undeniably uncertain. Firms, as market actors, can and must contribute to the net zero transition needed to stabilize climate, but the trajectory to that end state may include multiple dynamics associated with transitional changes in organizational life.

1.9   Unpacking Climate Risk as a Business Risk

Further insight into the portfolio of climate risks facing business, is found first by examining the definitions used to identify and classify climate risks, usually defined by risk governance and practitioners. Business definitions of climate risk have gone mainstream since they were first produced by the TCFD in its 2017 final report of recommendations for how companies should analyze and report on climate risk. The TCFD defines climate risk as having two distinct categories. The first category is that climate risk is a physical risk, associated with the physical manifestations of dangerous climate change impacts, either as an acute risk, i.e., sudden, extreme climate event, or as a chronic risk, i.e., the steady incremental global warming changes occurring over time.

The second category of climate risk defined by TCFD is that it is a transition risk associated with future regulatory, legal, market, and reputational risks, as consequents of a disorderly economic transition to a net zero economy. Both categorizations of risk and the implications for business are explained next.

1.10   PHYSICAL RISK

In the first instance, physical risk is created by climate events, or climate drivers. In practitioner language, drivers are risk sources or the events that create hazards. In the current context, climate drivers are the climate events which create physical hazards to business property and other assets in both the built and natural environment. According to climate modeling and IPCC climate assessment work, both acute and chronic physical risks are forecast to increase (Reisinger et al., 2020). Climate drivers such as rising temperatures and catastrophic climate events vary across climate zones and can be distinguished by their characteristics. For example, in some parts of the United States and Canada, catastrophic flooding and coastal sea level rise are viewed as climate drivers. In India and Asia, rising land temperatures and hurricanes have been recorded as more prevalent types of catastrophic climate drivers. In 2017 natural disasters attributed to climate change caused an estimated $306 Bn in economic damage globally, equivalent to the GDP for Denmark or Egypt (Swiss Re Institute, 2018).

To reduce the loss exposures arising from the hazards created by climate drivers, risk transfer mechanisms, i.e., property and casualty insurance, may offer partial short-term financial protection for corporate entities. However, the increase in catastrophic claims across firms and sectors is showing effects of 'insurance market hardening' producing cost prohibitive insurance products. This has the effect of producing a higher cost of risk for corporate groups. Even risk backstopping measures through extraordinary reinsurance coverages may not in the long run be able to underwrite property damage and business interruption claims as a result of acute weather events (McKinsey Global Institute, 2020). This scenario alters corporate risk appetites for re-enforcing business continuity, due to reductions in viable risk transfer and residual risk retention methods. In that extreme scenario, both insured and insuring parties can no longer carry the 'cost' of risk, either by the insured party, i.e., the firm, and the insurer, the primary and reinsurance providers.

1.11   TRANSITION RISK

Transition risk is the second formal category of climate risk applied to business. Transition risk is manifested as the risks associated with a disorderly, that is, an unmanageable transition to net zero future economy where exposures to regulatory, legal, market, and reputation risk exposures may occur. Corporate entities operating under new and stricter risk regimes, will need to contend with avoiding loss exposures associated with missed decarbonization targets. Corporate roadmaps targeting emission level reductions incrementally and accumulatively per year, are symbolic and aim to express corporate ambition. Where carbon performance falls short, corporate reputations reflected in stakeholder and investor responses, may be significantly affected. At the time of writing, carbon delay or inaction on the part of business is such a concern, some emissions registry groups have established transition accountability frameworks. CDP registered companies for instance are asked to voluntarily report on incremental decarbonization progress plans, under the CDP's Assessing Carbon Transition program (ACT) (Carbon Disclosure Project, 2021).

Other scenarios of transition risk are expected. Business reputation loss, liability exposures, regulatory exposures, and financial and investor exposures may result in divestiture, exchange de-listing, and business exit. None of this is desirable, but the unimaginable downfall of traditionally robust profit-taking companies is a realistic scenario to reckon with under extraordinary circumstances. Capital market preference for low carbon portfolios will likely grow, possibly in a winner take loser, zero sum game for corporate survival. Another scenario might envision protracted and unruly business transitions to accommodate dual objectives of risk prevention and corporate profit and growth. A more preferred scenario of course is one where organizations exemplify the behaviors and acquire the resources required to manage the transition that climate risk will bring to the organizational landscape. The graphical representation of climate-related risks and financial impact is provided in Fig. 1.1.

Another issue is the emergence of double materiality and the new risk model, discussed next.

1.12   DOUBLE MATERIALITY AND THE NEW RISK MODEL

The inclusion of climate-related disclosures will also have a profound effect on the way materiality is viewed within a company and by investor and stakeholder audiences. The nature of materiality, or what is understood as material or important to report (because the issue, i.e., the risk, will have financial implications) creates a new holistic risk model for business. Two well-known sustainability reporting frameworks illustrate

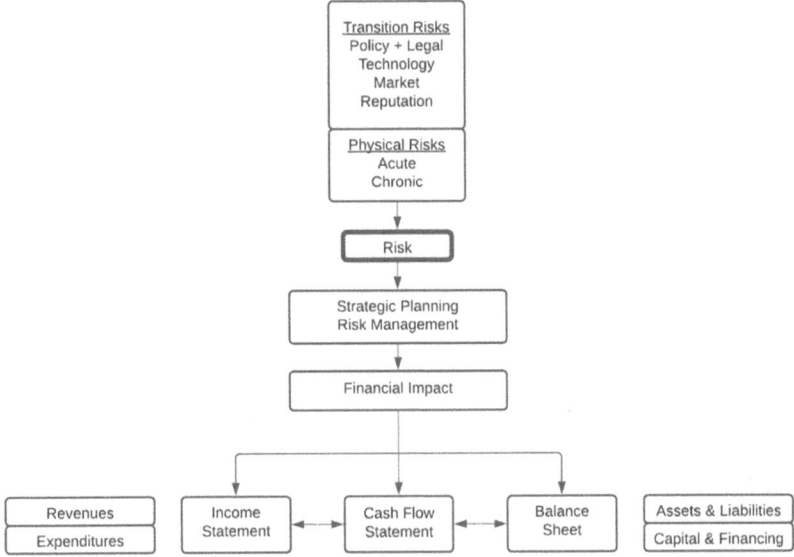

Fig. 1.1 Climate-related risks and financial impact (By Author, partially derived from TCFD 2017 Final Report on Climate Disclosures)

different reporting perspectives on the questions 'Material to whom?' and secondly, 'What effect does materiality have on the climate risk model?'.

Until recently, the non-financial corporate reporting groups, the Global Reporting Initiative (GRI) and the Sustainability Accounting Standards Board (SASB) have different meanings of materiality. The SASB definition of materiality was understood as how a company identifies and assesses those sustainability issues that influence enterprise value, or the 'financial materiality' or 'outside in impact'. On the other hand, the GRI definition of materiality was defined by the way a company identifies and assesses business impacts on the economy, environment, and people. This referred to environmental social and governance materiality (ESG) based on inside-out impact of corporate behavior. Efforts to integrate or create interoperability between these perspectives to better reflect risk exposures are of ongoing concern for reporting groups and business practitioners. In literature, double materiality of climate risk was recognized in Gasbarro and Pinkse (2013) where they assert '[w]hile business tends to be seen as a substantial factor in causing climate change, climate induced physical

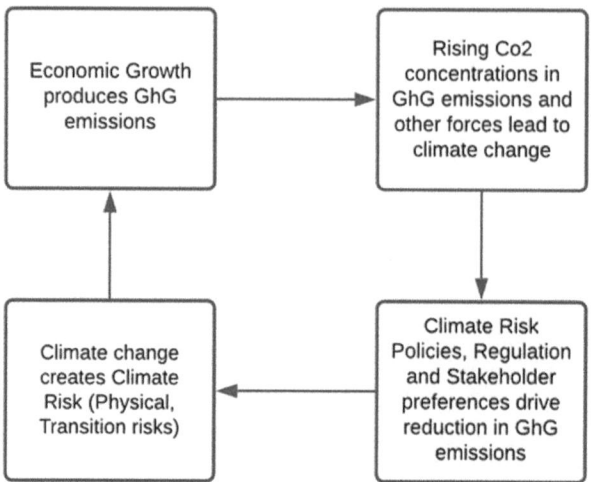

Fig. 1.2 New risk model of double materiality (By Author)

changes can also pose major challenges to firms in return' (Gasbarro & Pinkse, 2013, p. 179).

As an abstraction, the climate risk model can be seen as the representation of a system that locates risk sources both within exogenous and endogenous locations in the business environment. For graphical illustration, the new risk model of double materiality as being both an impacts-out and an impacts-in phenomena, is depicted in Fig. 1.2.

1.13   Production of Radical Business Challenges

As discussed, climate risk is created by the acute and chronic hazards associated with broad physical climate change impacts, and the secondary, systemic, and localized risk regime exposures to regulation, legal, financial, and reputation losses. Physical and transition risk exposures require companies to activate response actions of rapid decarbonization across core operations, supply chains, and customer products. They must do this to remain in alignment with home country climate risk disclosure policies and regulations.

In view of the above phenomena, I tentatively proffer resulting business challenges involve managing the low carbon initiatives, and will have varied, vast, and complex effects. Transitions of the order I am discussing, defy prior business assumptions about organizational capacity at the cognitive, resource, and risk management levels of the organization. Challenges exist at every level of organizational life, most tangibly noted in risk management practices that do not adequately manage the new risk category of climate risk.

This is concerning given deep uncertainty of future climate events and consequential economic events. Corporate disclosures and reporting, including the articulation of future business strategies in a decarbonized economy imply unprecedented planning horizons, possibly unworkable to many corporate managers. Corporate disclosures providing assurances to investors and stakeholders of decarbonization progress, will require new skills and expertise, external due diligence of GhG metrics, innovations in financing low carbon initiatives, and a spectrum of intra organizational changes. These are discussed in the upcoming chapters.

1.14 REFLECTION

So now we arrive at a point of reflection about the choice businesses have under anticipated and increasingly stringent climate risk regime rules. Corporate leaders and governance groups can motivate and guide their organizations through protracted decarbonization to meet 2050 net zero targets. In the process, and if done in sufficiency, the firm achieves the strategic objective of building out longer term mitigative capacity for the firm. Alternatively, and undesirably, it could face risk of business failure in the long run. That is the 'what' of the issue. While the choice may seem straight forward, how business responds to what is conceptually an existential threat to its own survival, is far from a simple matter. The implications of managing climate risk exposures to protect and ensure business continuity in a low carbon economy, are varied, vast, and complex. Climate risk is a future-bound construct, and in the strictest sense of the term, it does not exist, yet. Incipient business practices to reduce expected climate risks are underdeveloped. The future ahead is unresolved and saturated with speculation. Limited empirical evidence exists to test business theory through exemplary (and non-exemplary) corporate behavior because the full force of climate risk has not (fully) happened yet. Hence, one cannot quite contemplate the full range of dynamics of what a business decarbonization transition will look like. Optimists and historians remind us that history says 'capitalism always finds a way' to solve problems. What can also be done here with this book is demonstrate how theoretical speculations found in literature offer at least a partially useful explanation for how business will likely manage the set of climate risk challenges. In multiple examples from theoretical and empirical work, this book also addresses how business carbon reduction transitions are likely to alter theoretical landscapes, especially as they apply to behavioral, resource, and risk disclosure theories generated over the last 80 years of scholarship. Arguable is that the current requirement of corporations to 'solve the problem they created' (Wright & Nyberg, 2017) will render extant business and management theories and practices either inadequate or in need of revision.

1.15   Purpose of This Book

The purpose of this book is threefold.

The first purpose is to examine business challenges created by climate risk through an explicit risk management lens. Framing climate risk as business risk not only reflects the current real-world expectations of stakeholders and financial audiences. It also echoes the growing shift in climate scholarship to accommodate organizational climate responses as risk based.

1.16   Risk Mitigation as Response

The climate risk mitigation response is the dominant risk construct discussed in this work, within the application domains of management cognition and risk beliefs, organizational resources, informal and formal risk management practices. Deliberately excluded are discussions pertaining to (1) the opportunities of climate risk, (2) mitigation strategies involving geoengineering and global scale technology, and (3) mitigation through emissions trading. This is done to preserve discussion boundaries and to assist in keeping the enormity of the topic to a manageable scale.

Risk mitigation responses can be viewed as fundamentally both the same and different across different business sectors, i.e., consumer, financial, industrial, energy, and different classifications of businesses, i.e., reporting companies, privately held firms, and hybrid organizations. Unifying themes and approaches are offered to streamline the discussion and to encourage useful and reflective thinking about decarbonization transition, in a direction that assists academics and practitioners.

The second purpose is to examine relevant and extant theories in a way that demonstrates where academic research can contribute to better understanding of business transition challenges. Part of that aim will be to problematize extant theories by demonstrating their limitations to account for present day mitigation efforts. In the contrarian paraphrased words of Karl Popper in 1959, 'Do theories account for all their implications and do they consider possible unintended consequences in depth'?

The third purpose of this book is to suggest research questions and opportunities throughout, that are alive to the unique conditions businesses face. As a result, it is my aim to shed light will be shed on how firm-level response can contribute to global climate mitigation goals.

1.17   INTERDISCIPLINARY LITERATURE STREAMS

Climate risk is a complex subject spanning disciplines from climate science, economics, behavioral theory, organizational and risk management. Literatures used in this work are drawn from a constellation of sources—from special interest areas in business and management literature, risk theory and risk practitioner literature, financial, accounting and sustainability literature—as well as cross domain research in behavioral economic, cognitive science, and psychology. Both empirical and theoretical contributions are included, with notable emphasis on literature published after the Paris Agreement in 2015. It is my view that management literature tends to reflect on its own special interest domain and concentrate less so on the contributions of other fields. While this may be beneficial to knowledge gap analysis within a certain domain, it also results in siloed thinking that discounts other domain contributions that would otherwise provide insight and benefits. This is the perennial argument in academe in favor of interdisciplinary studies, of which I am a proponent. Linking approaches from one domain to another can produce novel and fruitful solutions to systemic management problems, and I propose that the topic of climate risk and business is one such application of this idea.

At the time of writing, there remains relatively limited qualitative and even less quantitative robust empirical evidence of ongoing decarbonization effort in industry, with exceptions. I have endeavored to reference as many of them as word counts dictate in the chapters ahead.

As a result, three key points are to be made.

First, the knowledge base in literature of climate risk management by the private sector, is limited. To date, organizational response literature on shorter term adaptation responses and resiliency of organizations to withstand climate impacts has been well addressed (Linnenlueke, 2015). Much work in adaptation is not framed as a risk response but as an outcome of organizational change processes. Some exceptions include Obersteiner et al. (2001) on 'Managing Climate Risk' and Weinhofer and Busch's (2013) 'Corporate Strategies for Managing Climate Risk,' both of which conceptualize the longer term climate impacts as distinct business risk. However, since 2015 and the Paris Agreement, and the 2017 release of the TCFD guidelines for corporate climate risk reporting, more contributions on climate risk management have appeared in literature.

My own field study work on risk cognitions of chief executive officers in the electricity sector illuminated an unexpected management preference for firm-level resources and capabilities to manage future climate risk (Dowbiggin, 2018).

Second, theoretical and empirical coverage of the use of new climate risk reporting metrics and tools promulgated by reporting groups, such as the science-based targets initiative of the CDP and TCFD reporting guidelines, still remains limited in management, accounting, and financial literatures. Noted by O'Dwyer and Unerman (2020), and Bebbington and Unerman (2018), it is an issue of concern in sustainability accounting literature, compelling O'Dwyer and Unerman (2020) to assert 'There has been little, if any, substantive academic research published on this potentially transformative corporate reporting initiative (of the TCFD)' (p. 1133).

Third, in practice, decarbonization ambitions are gradually accelerating, as evidenced by the increasing number of corporate registrations to the Science Based Targets initiative (SBTi) of the CDP. However, current levels of engagement suggest actual response actions by companies are still underdeveloped. As such, this has produced a noticeable deficit in the literature, possibly a result of low empirics, low awareness, prolonged timelines for publication, or a 'wait and see' approach by observers. It is on this third point that I suggest to readers that there is no time to waste in examining the hypothetical challenges of climate risk for business. Considering the limited empirics of the phenomenona, I suggest an appropriate approach to this topic is one that is aligned with the concept of a new thought experiment, discussed next.

1.18 New Thought Experiment

Given the limitations associated with exploring novel hypothetical climate risk mitigation challenges for business, discussions in this book are provided through deliberations that speculate about the consequences of climate risk management for organizations. This is done through the new thought experiment approach in this book. To be clear, thought experiments are the enactment of managing novel hypothetical situations and thinking through their consequences. This approach is the hallmark of a thought experiment, aligning with Yeates' (2004) definition: 'A thought experiment is a device with which one performs an intentional, structured process of intellectual deliberation to speculate, within a specifiable problem domain, about potential consequents (or antecedents) for a designated antecedent or consequent' (p. 150).

Thought experiments are also a form of philosophy, and this can be useful in this research context where norms and values are concerned. Thought experiments are not a replacement for empirical work nor technically a 'real' experiment, but a complementary force to scholarly work. One enduring definition of thought experiments was proposed by Kuhn (1977), who argued: 'Thought experiments give the scientist access to information which is simultaneously at hand and yet somehow inaccessible to him' (Kuhn, 1977, p. 261). In more recent work, distinguished business scholars have actively used thought experiments in published works by looking at empirical data and evidence from different and unusual perspectives. Examples of thought experiments are found in published works by Prahalad and Hamel (1990) and Porter (1989). In doing so, they seized 'moments of crises' induced by the failure of theoretical expectations and then cast light on the meanings of ideas not fully grasped to 'visible tensions, contradictions and anomalies that are the fuel for new theorizing' (Kornberger & Mantere, 2020, p. 5).

This book aims to function as a thought experiment by design 'that shows something, to reveal something that is of interest yet has been neglected or simply overlooked. One can easily grasp the potential implications for organization theory; if thought experiments generate, demonstrate, and communicate new ideas, then organization theory would be well served to develop thought experiments. It would provide valuable heuristic insights and allow communication of its core ideas to a non-expert audience. Such work should not replace more traditional scholarly work, but rather be complementary to it' (Kornberger & Mantere, 2020, pp. 5–6).

1.19   Organization of This Book

While the overall goal of this book is to present discussions about the challenges facing business in such a way to encourage useful and reflective thinking, I present these discussions through alternate frames, in a tentative fivefold conceptual framework in the following chapters. The framework is constructed along five dimensions of transition, reminiscent of Geels (2002).

I discuss the five dimensions next.

1.20   Cognitive Challenges

The first business challenge discussed ahead in Chapter 2 is one of cognitive transitions created by shifts in management cognition and risk belief. The emergence of climate risk appearing on multiple fronts in business implies a need for different ways of regarding risk in organizations, resulting in what theorists call a 'cognitive transformation in a complex setting' (Wittneben et al., 2012). The attainment of higher order complexity thinking is associated with elevated management cognition, and risk beliefs by company actors that manifest themselves as risk identifiers in new and non-traditional ways. Extant literature in these areas is paltry however, as it underserves what critics call an emerging management crisis to recognize and undertake the work needed to address the climate crisis (Wiltshire et al., 2014). Climate risk complicates business in many ways, starting with how decision-makers think of future risks beyond their personal experience and current knowledge of systemic exposure and vulnerability. As well, lack of attention to a particular risk is often driven by a belief that all risks have been identified. Low cognitive processing of risk, or reluctance to perceive the importance of prioritizing one set of risks over another, may limit and possibly amplify unattended risks, leading to negative outcomes. This is discussed with several objectives in mind, including illustrating from prior work how cognition and risk belief enable firm-level action and sustain transition processes, and to show where extant theories could be enlivened to address climate risk perceptions at the firm level.

1.21   Resource Challenges

The second business challenge discussed ahead in Chapter 3 addresses how firm-level resources will undergo transition. Literature already suggests that current organizational resources, processes, and routines are not well equipped to support massive transformational change in organizations, even in the face of win–win, competitive advantage objectives (Beer & Nohria, 2000; Burnes, 2004; Mellahi & Wilkinson, 2010; Styhre, 2002). Organizational resource transition driven by 'loss' scenarios, will prove interesting, if prospect theory is held to account (Kahneman & Tversky, 1979). Which organizational resources and other internal factors result in the production of mitigative capacity arises as an important consideration.

1.22   Risk Practice Transition—Informal and Formal

The third and fourth business challenges discussed in Chapters 4–6 involve how firms will moderate informal and formal risk practices. Current risk management practices in industry may not adequately integrate climate risk, given deep uncertainty and nonlinearity of climate-related impacts, the specter of time horizons beyond normal risk forecasting practices, and the complications of analyzing risks that, if materialized, produce multiple future states for the organization.

I proffer that transition will alter risk culture, risk translation, and produce innovation in risk management practices. Under TCFD guidelines, firms must describe how their process for identifying, assessing, and managing climate-related risks are integrated into their overall risk management systems. While conventional risk management processes are designed to inoculate uncertainty and complexity out of risk assessment, 'counting and calculating' climate risk adds an additional complexity driven by lack of risk controls. How climate risk shifts traditional enterprise risk management systems away from rules-based management to an approach which addresses the 'critical management of alternative futures' is a case in point (O'Dwyer & Unerman, 2020).

1.23   Materiality and Disclosure Challenges

The fifth business challenge discussed in Chapter 6, is an extension of Chapter 4 and 5 materials, and addresses the shifts and new requirements of corporate climate risk disclosure and reporting under TCFD reporting. Those challenges include understanding new TCFD mandated sustainability reporting practices, accounting for sector-specific materiality, and accommodating competing reporting frameworks, i.e., ESG, IIR, for different reporting audiences. This fifth challenge is concerned with the transformative reporting requirements for companies to address climate-related materiality determination, and other shifts in sustainability reporting practices. The magnitude of this challenge is lightly understood as evidenced by low levels of academic papers on this topic. The shift from impact reporting to risk and dependencies reporting for companies has implications for the development of corporate capacity and abilities. Corporate disclosure and shareholder reporting regarding GhG emissions levels across three scope zones of the business value chain, will require new standards of measurement, interpretation, and assurance. Companies may have relative strategic control over their supply chain and its own direct emissions levels, but unable to carbon-account for GhG levels produced by its downstream products and services, i.e., Scope 3 emissions. Due diligence of a company's efforts to decarbonize in alignment with science-based targets established by the SBTi initiative of the Carbon Disclosure Project for example, will call into question the reporting credibility and reputation of the firm, creating new areas of non-tangible asset exposures. Complex corporate reporting requirements for business model dependencies based on extraordinary 30–60-year time horizons may seem unworkable to corporate managers. A simplified model of business challenges addressed in this work, is provided below (Fig. 1.3).

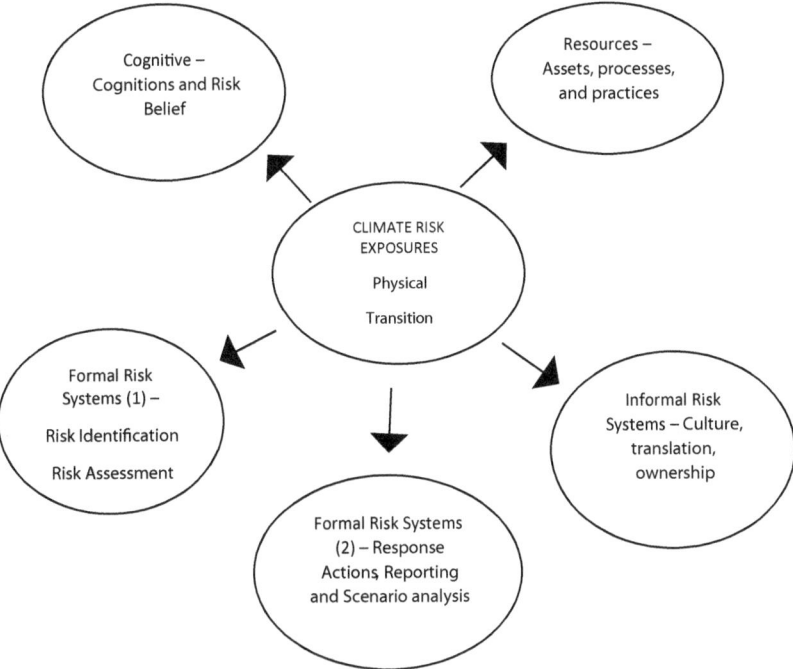

Fig. 1.3 Firm-level transition challenges of climate risk

REFERENCES

Allen, M. R., Frame, J., Huntingford, C., Jones, C., Lowe, J., Meinshausen, M., & Meinhausen, N. (2009). Warming caused by cumulative carbon emissions towards the trillionth tonne. *Nature, 458*, 1163–1166. https://doi.org/10.1038/nature08019

Bebbington, J., & Unerman, J. (2018). Achieving the United Nations Sustainable Development Goals: An enabling role for accounting research. *Accounting, Auditing & Accountability Journal, 31*(1), 2–24.

Beer, M., & Nohria, N. (2000). Cracking the code of change. *Harvard Business Review, 78*(3), 133–141.

Bui, B., & De Villiers, C. (2017). Business strategies and management accounting in response to climate change risk exposure and regulatory uncertainty. *The British Accounting Review, 49*(1), 4–24. https://doi.org/10.1016/j.bar.2016.10.006

Burnes, B. (2004). Kurt Lewin and the planned approach to change: A re-appraisal. *Journal of Management Studies, 41*(6), 977–1002. https://doi.org/10.1111/j.1467-6486.2004.00463.x

Cadez, S., & Czerny, A. (2016). Climate change mitigation strategies in carbon-intensive firms. *Journal of Cleaner Production, 112*, 4132–4143. https://doi.org/10.1016/j.jclepro.2015.07.099

Carney, M. (2021). *Value(s): Building a better world for all.* Penguin Random House Canada.

Carbon Disclosure Project. (2021). *Assessing Carbon Transition.* https://actinitiative.org/

Carbon Trust. (2021). *Briefing: What are Scope 3 emissions?* https://www.carbontrust.com/resources/briefing-what-are-scope-3-emissions

de Sousa Jabbour, A. B. L., Chiappetta Jabbour, C. J., Sarkis, J., Gunasekaran, A., Furlan Matos Alves, M. W., & Ribeiro, D. A. (2019). Decarbonization of operations management–looking back, moving forward: a review and implications for the production research community. *International Journal of Production Research, 57*(15–16), 4743–4765. https://doi.org/10.1080/00207543.2017.1421790

Dowbiggin, A. (2018). *Climate risk perceptions in the Ontario (Canada) electricity sector* (Doctoral Dissertation). Heriot Watt University.

Flusberg, S. J., Matlock, T., & Thibodeau, P. H. (2017). Metaphors for the war (or race) against climate change. *Environmental Communication, 11*(6), 769–783. https://doi.org/10.1080/17524032.2017.1289111

Gasbarro, F., & Pinkse, J. (2013). Managing physical impacts of climate change: How awareness and vulnerability induce adaptation. *Academy of Management Proceedings, 2013*(1), 11870. https://doi.org/10.5465/ambpp.2013.11870abstract

Geels, F. W. (2002). Technological transitions as evolutionary reconfiguration processes: A multilevel perspective and case study. *Research Policy, 31*(8/9), 1257–1274.

Intergovernmental Panel on Climate Change. (2014). *Climate Change 2014: Mitigation of Climate Change.* https://www.ipcc.ch/site/assets/uploads/2018/02/ipcc_wg3_ar5_full.pdf

Intergovernmental Panel on Climate Change. (2021). *Working Group III Mitigation of Climate Change.* https://www.ipcc.ch/working-group/wg3/

International Energy Agency. (2019). *Global Energy Review 2019.* https://www.iea.org/reports/global-energy-review-2019

Kahneman, D., & Tversky, A. (1979). Prospect theory: An analysis of decision under risk. *Econometrician, 47*(2), 263–292. https://doi.org/10.2307/1914185

Kolk, A., Levy, D., & Pinkse, J. (2008). Corporate responses in an emerging climate regime: The institutionalization and commensuration of carbon disclosure. *European Accounting Review, 17*(4), 719–745.

Kornberger, M., & Mantere, S. (2020). Thought experiments and philosophy in organizational research. *Organization Theory, 1*(3), 1–19. https://doi.org/10.1177/2631787720942524

Kuhn, T. (1977). *The essential tension: Selected studies in scientific tradition and change*. The University of Chicago Press.

Lee, S. M., Kim, S. T., & Choi, D. (2012). Green supply chain management and organizational performance. *Industrial Management & Data Systems, 112*(8), 1148–1180. https://doi.org/10.1108/02635571211264609

Linnenlueke, M. (2015). Resilience in business and management research: A review of influential publications and a research agenda. *International Journal of Management Reviews, 19*(1), 4–30. https://doi.org/10.1111/ijmr.12076

McKinsey Global Institute. (2020). *Climate risk and response: Physical hazards and socioeconomic impacts*. https://www.mckinsey.com/business-functions/sustainability/our-insights/climate-risk-and-response-physical-hazards-and-socioeconomic-impacts

Mellahi, K., & Wilkinson, A. (2010). Managing and coping with organizational failure: Introduction to the special issue. *Group & Organization Management, 35*(5), 531–541. https://doi.org/10.1108/AAAJ-02-2020-4445

Nordhaus, W. (2013). *Climate Casino: Risk, uncertainty, and economics for a warming world*. Yale University Press.

Obersteiner, M. (2001). *Managing climate risk*. International Institute for Applied Systems Analysis. http://pure.iiasa.ac.at/id/eprint/6471/1/IR-01-051.pdf

O'Dwyer, B., & Unerman, J. (2020). Shifting the focus of sustainability accounting from impacts to risks and dependencies: Researching the transformative potential of TCFD reporting. *Accounting, Auditing & Accountability Journal, 33*(5), 1113–1141. https://doi.org/10.1108/AAAJ-02-2020-4445

Porter, M. E. (1989). How competitive forces shape strategy. *Harvard Business Review, 2*, 137–145.

Prahalad, C. K., & Hamel, G. (1990). The core competence of corporation. *The Harvard Business Review, 33*, 79–91.

Pye, S., Broad, O., Bataille, C., Brockway, P., Daly, H. E., Freeman, R., Gambhir, A., Geden, O., Rogan, F., Sanghvi, S., Tomei, J., Borushylo, I., Watson, J. (2021). Modelling net-zero emissions energy systems requires a change in approach. *Climate Policy, 21*(2). https://doi.org/10.1080/14693062.2020.1824891

Reisinger, Andy, Howden, M., Vera, C., Garschagen, M., Hurlbert, M., Krei-biehl, S., Mach, K., Mintenbeck, K., O'Neill, B., Pathak, P., Pedace, R., Pörtner, H., Poloczanska, E., Corradi, M., Sillmann., J., van Aalst, M., Viner,

D., Jones, R., Ruane, A., Ranasinghe, R. (2020). *The Concept of Risk in the IPCC Sixth Assessment Report: A Summary of Cross-Working Group Discussions*. Intergovernmental Panel on Climate Change. https://www.ipcc.ch/site/assets/uploads/2021/02/Risk-guidance-FINAL_15Feb2021.pdf

Stenhouse, N., Maibach, E., Cobb, S., Ban, R., Bleistein, A., Croft, P., ... & Leiserowitz, A. (2014). Meteorologists' views about global warming: A survey of American Meteorological Society professional members. *Bulletin of the American Meteorological Society, 95*(7), 1029–1040.

Styhre, A. (2002). Non-linear change in organizations: Organization change management informed by complexity theory. *Leadership & Organization Development Journal, 23*(6), 343–351. https://doi.org/10.1108/01437730210441300

Swiss Re Institute. (2018). *Natural catastrophes and man-made disasters in 2017: A year of record-breaking losses.* https://reliefweb.int/sites/reliefweb.int/files/resources/sigma1_2018_en.pdf

TCFD. (2017). *Recommendations of the Task Force on Climate-related Financial Disclosures.* https://assets.bbhub.io/company/sites/60/2020/10/FINAL-2017-TCFD-Report-11052018.pdf

TCFD. (2020). *2020 Status Report.* https://www.fsb.org/wp-content/uploads/P291020-1.pdf

The White House. (2017). *Presidential executive order on promoting energy independence and economic growth.* https://unfccc.int/files/focus/long-term_strategies/application/pdf/mid_century_strategy_report-final_red.pdf

Weinhofer, G., & Busch, T. (2013). Corporate strategies for managing climate risks. *Business Strategy and the Environment, 22*(2), 121–144. https://doi.org/10.1002/bse.1744

Wiltshire, T. J., Neville, K. J., Lauth, M. R., Rinkinen, C., & Ramirez, L. F. (2014). Applications of cognitive transformation theory: Examining the role of sensemaking in the instruction of air traffic control students. *Journal of Cognitive Engineering and Decision Making, 8*(3), 219–247. https://doi.org/10.1177/1555343414532470

Wittneben, B. B., Okereke, C., Banerjee, S. B., & Levy, D. L. (2012). Climate change and the emergence of new organizational landscapes. *Organization Studies, 33*(11), 1431–1450. https://doi.org/10.1177/0170840612464612

Wright, C., & Nyberg, D. (2017). An inconvenient truth: How organizations translate climate change into business as usual. *Academy of Management Journal, 60*(5), 1633–1661. https://doi.org/10.5465/amj.2015.0718

Yeates, L. (2004). *Thought Experimentation: A Cognitive Approach* (Doctoral Dissertation). University of South Wales, 2002.

Zizzo, L. (2017, June 12). Companies can no longer hide their climate risk. *The Globe and Mail.* https://www.theglobeandmail.com/report-on-business/rob-commentary/companies-can-no-longer-hide-their-climate-risk/article35288626/

CHAPTER 2

Cognitive Challenges

Abstract The foremost underlying force for all business response actions to climate risk lies at the level of cognition among and between organizational actors. Cognition, or the way in which company actors think about climate risk in its various dimensions, drives response actions to mitigate climate risk, guided by two assumptions. One, firms will be dealing with a novel set of 'deeply uncertain' risks previously unconnected with organizational resources, processes, and informal and formal risk practices. Two, firms will be enacting risk response preferences through private acts of corporate decision-making (later to become publicly disclosed in most cases) based on their cognition of the response need, their interpretations of reliable data and information, and their perception of the risk features attributed to climate risk.

Keywords Management cognition · Perception · Risk belief · Cognitive transition · Cognitive complexity · Cognitive mastery · Cognitive fusion · Ontological standardization · Mental modeling

25

2.1   INTRODUCTION

The introduction of new risk categories to an organization represents a major transition in organizational priorities and management practices. The challenges of climate risk to business will most likely be varied, vast, and complex. As proposed at the outset of this book, I proffer a tentative fivefold proposition about how climate risk will theoretically challenge business at multiple levels of organizational life. Reflecting on probable corporate response actions to reduce climate risk, invites the question of what the consequences will be for organizational cognition, resources, and informal and formal risk practices.

2.2   CONSEQUENCES OF CLIMATE RISK

Organizations are likely to be severely impacted by climate risk in combinative ways, depending on which climate risk is under analysis. Climate risks that are physical in nature, i.e., sea level rises, flooding, and other extreme weather events—deemed as 'acute' risks, or the gradual increase in temperature, deemed as a 'chronic' risk, represent near and long-term loss exposures to infrastructure, facilities, assets, labor shortages, and business continuity. Climate risks which are transitional in nature—deemed as financial, regulatory, policy, technology, and legal liabilities, potentiate loss exposures associated with a 'disorderly transition' to a low carbon economic world (Battiston et al., 2021).

Both physical and transitional types of climate risks are equally and potentially harm-producing in cases where companies do not develop sufficient response capacity to deal with climate-related risks. The above scenarios implicate business in a vexatious range of transition challenges with far-reaching effects at the cognitive, organizational, and strategic level of the firm. As stated, it is the goal of this book to tentatively propose what business can expect to occur, using a firm level of analysis. Useful new thinking may emerge in this thought experiment, not only to suggest requirements for new empirical work, but how prior theoretical contributions may now be in need of revision. Academic research can contribute toward new and useful approaches of understanding how and why business response actions involve changes in cognitions, resources, and informal and formal risk management practices. These are the elements in the fivefold proposition introduced in Chapter 1.

2.3   Organization of This Chapter

In this chapter, I posit the foremost underlying force for all business response actions to climate risk, lies at the level of cognition among and between organizational actors. Cognition, or the way in which company actors think about climate risk in its various dimensions, drive response actions to mitigate climate risk, guided by two assumptions:

One, firms will be dealing with a novel set of 'deeply uncertain' risks previously unconnected with organizational resources, processes, and informal and formal risk practices. Two, firms will be enacting risk response preferences through private acts of corporate decision-making (later to become publically disclosed in most cases) based on their cognition of the response need, interpretations of reliable data and information, and the perception of the risk features attributed to climate risk.

I propose that the tangible aspects of organizational life, i.e., resources, processes, and risk practices discussed ahead in Chapters 3–6 will not necessarily be enacted upon unless there is an underlying cognitive basis for believing those tangible aspects are necessary in the first place. In this respect cognition plays a principal underlying role to how firms handle and reduce climate risk.

Given these assumptions, the role cognition plays in the debate surrounding business and climate risk challenges is, as I suggest, manifested in two ways:

- Cognitive mastery over the complexities imparted by climate risk.
- Cognitive fusion driven by ontological standardizations of climate risk.

In the organization of this chapter, a preliminary overview is first made of cognition literature leading up to linkages with climate risk. The overview demonstrates multiple perspectives as well as how far extant cognition literature can be held to account for today's business challenges with climate. Necessary to the discussion of cognition are the related topics of how epistemology, ontology, and semantic-linguistic manipulations can alter cognitions and risk belief systems within organizations. The overview is followed by my tentative proposition that outcome effects on management cognition occur also as a result of (a) climate complexity and (b) forced ontological standardization of climate risk by institutional groups. I start first with the literature on cognition in business.

2.4   COGNITION IN BUSINESS

Cognition is an intangible aspect of business management, but its recognition has been studied in management, as well as cognitive science and social psychology literatures for over 70 years. The genesis of the topic as an antecedent force in organizational life grew from earlier propositions that management cognition was objectivist, rationally bounded, and intellectually homogeneous (Simon, 1955). Organizational behavior was viewed as an outcome of utility seeking individuals producing the same or similar responses to management challenges (Dowbiggin, 2018). Evolving theoretical perspectives contested this notion claiming that differentiated organizational strategy and the available resources and structures in the company, had much more to do with company performance (Chandler, 1962; Stubbart, 1989). Chandler's 1962 strategic management proposition diverged from that notion claiming strategic planning and control comes about from 'conscious human deliberation', implicitly suggesting a cognitive basis for much of the strategy-making process (Chandler, 1962).

In further theory development, Stubbart's (1989) work drew greater attention to management cognition theory and somewhat stridently, admonished rationality purported by Simon (1955) as 'an ideal rather than as an empirical fact' (Smircich & Stubbart, 1985, p. 238). Stubbart's view was that business environments were 'enacted' or formed through social construction and interactive processes of constituent groups, giving rise to subjective views of decision-making.

Other work imported from social psychology domains proposed that corporate performance could be explained via social cognitive theory—where management cognition is heavily influenced, for example, by individual ideas of managerial efficacy beliefs, goal setting and the quality of analytical thinking (Wood et al., 1987). In that work, Wood et al. were hinting at a cognition—belief connection, which, if induced or manipulated could achieve certain outcomes such as judgment, attribution, and decision, especially in situations where the individual has a problem to solve.

In individuals tasked with problem-solving, the organization of conceptually related representations of objects, situations, and sequences of events and actions, is created as a cognitive structure. Cognitive structures are generically understood as a schema or as a personal construct (Kelly, 1955), or as a mental map of how the individual construes concepts and relationships between concepts (Weick, 1979). Exploration of the mental map of individuals within an organization helps organizations understand why individuals behave as they do. The interconnection between antecedent cognitive framing of issues and consequent behaviors and actions is a generally accepted axiom in cognitive psychology, explicated by two guiding principles that: (a) cognition that is context specific and directly associated with the environment of the individual/organization; and that (b) the perception of environmental objects, situations, and sequences of events and actions create cognition, which in turn effects behavior i.e., action, ultimately producing outcomes resulting from actions.

I suggest that these earlier theoretical anchors found in social-psychological theory reasonably account for decision-making based on process-based cognition, and by 'how the situation, stimuli and the variables controlling the responses are represented in the minds of the participants' (Markus & Zajonc, 1985, p. 137).

Perspectives given by social psychology theorists were a divergence in cognition theory from objectivist views to the recognition of individual 'sovereignty of subjectivity' then to the 'permeability of cognition' through manipulations relating to belief systems, language framing and interpretation. This has led management scholars such as Sharma (2000) to illuminate the link between cognition and managerial decision-making as a function of organizational interpretations.

Similarly, and put ever so more directly, risk governance scholar Sidortsov (2014) asserts that the level and quality of risk cognition, or the way corporate actors think about risk, drives what is actually done about it.

While cognition-performance linkages in literature have been well studied, other researchers, attentive to the climate reality, have begun to address the cognition-climate risk linkage (i.e., Dowbiggin, 2018; Gasbarro & Pinkse, 2015; Weinhofer & Busch, 2013) from the perspectives of risk perception, risk readiness and responses to institutional generated risks. Notable among them is a discernable paradigmatic 'splitting' represented by the tension between corporate objectives of risk taking and growth ensconced in the competitive advantage paradigm, to organizational objectives of risk reduction and caution. This theoretical tension between competitive advantage and climate risk reduction paradigms may well beacon a paradigmatic shift in the literature to accommodate current empirics on cognition.

2.5   RISK THEORY LITERATURE

In the current context of climate risk, cognition is inextricably connected and relevant to risk and risk belief constructs, as supported in literature.

Extant approaches in literature offer multiple perspectives on cognition and risk beliefs with prior work based on empirical study while other work based on theoretical speculations. Indeed, the contribution of risk theorists is considered by some to be lesser work for lack of empirical investigation (Sidortsov, 2014). Yet theoretical constructions are vital to study of business and climate risk challenges given the full extent of business response actions to climate risk is still unfolding—leaving few options other than theory to offer explanatory power. It can be noted that closing this theoretical gap would become an important source of enlivenment as more empirical evidence is analyzed and published in literature.

Understandings of risk and risk beliefs supported in risk literature from a variety of perspectives, are discussed next.

Risk theory literature covers multiple topics and subtopics both theoretically and empirically (Outhwaite, 1987). The corpus has grown noticeably since the 1970s, coinciding with growth in societal concern for nuclear reactor safety and the systemic consequences of natural disasters, e.g., the 1979 Three Mile Island nuclear reactor accident and the 2011 Fukushima tsunami in Japan, respectively (Bevere & Weigel, 2021). Present day insurance market 'hardening,' reinsurance market complexities and concerns for global financial stability represent recent risk applications in additional interest area domains in literature.

In literature, different risk theory schools emerged across disciplines. In North America, exemplary contributions in decision science emerged about risk perception and emotional affect (Slovic & Peters, 2006), and gain–loss prospect theory and decision-making under uncertainty (Kahneman & Tversky, 1979). Sociologist Ulrich Beck in the United Kingdom promulgates a risk-is-everywhere construct in his World Risk Society framework (Beck, 1999). This is by no means an exhaustive list of the risk authorship that supports a major interdisciplinary field of research. Risk contributions span various risk domains including operational cyber risk, global political and terrorism risk, and financial credit and market risks. Furthermore, because the field is sufficiently multidimensional it segments risk practice areas (e.g., business continuity, crisis management, risk financing, reinsurance accounting), and partitions journals by risk specialties, e.g., The Journal of Risk, Risk Analysis, Journal of Risk Research, Journal of Risk and Uncertainty, and the Journal of Risk and Insurance as some examples.

2.6   Cognition and Philosophical Treatments of Risk

Interpretations of the risk construct and its meaning are invigorated by the epistemology, ontology, and semantic-linguistic branches of philosophy. The epistemology of risk, or the knowledge component of risk; the ontology of risk, or how the truth of risk is regarded, and the semantic-linguistic use of the word risk, underscores how risk perception is assembled in cognitive processes.

2.7   Cognition and Epistemological Manipulation

To start, one could contemplate that the concept of risk as being epistemologically questionable, if not flawed altogether. Among other objectives, epistemological study is involved with how knowledge is regarded as the truth. In the case of risk, where uncertainty is unquestionably key to risk—the epistemological stance is called into question, raising the inevitable question 'why?.' The primary condition of risk is that of an emergent state associated with situations in which it is possible but not certain that some undesirable event will take place (Dowbiggin, 2018). In other words, where there is a risk, there must be something unknown about the situation, or the situation has an unknown outcome. Hence, knowledge about risk is knowledge about the lack of knowledge (Möller & Hansson, 2008). Next, cognition and ontology are discussed.

2.8   Cognition and Ontological Manipulation

Like all risks, climate risk has an ontological position that may vary with different sources and audiences. In earlier contributions to decision science, risk was viewed as a statistically expected loss (Wald, 1939) while Kaplan and Garrick (1981) defined risk in similar quantitative fashion as the likelihood and severity of events.

Social constructionist theorists such as Sandman (1987), designated risk as the subjective combination of hazard and outrage as did other risk theorists including Fischhoff et al. (1984) who asserted that risk definitions are a political act which expresses the definers' values regarding the relative importance of different possible adverse consequences for a particular decision.

Addressing climate risk, integral theorist Esbjorn-Hargens (2010) raises questions about the ontological status of climate change impacts as having multiple meanings to different audiences. Esbjorn-Hargens suggests that climate change impacts manifest as a multi-object phenomenon that is seen, revealed, and enacted in different disciplines. He asserts this ontological pluralism has multiple and different consequences for the different disciplines which 'enact' or reveal impacts related to climate risk.

By extension, the ontology of climate risk raises questions about how climate risk is 'enacted' by different groups given that climate risk cannot be 'seen' nor completely revealed in its entirety. In this regard, the ontological position of climate risk is relevant to the risk management techniques used by firms to categorize risks on the basis of materiality. Where standards and stringency of climate risk reporting now exceed those required in prior sustainability reporting the ontological position of climate risk becomes salient. Inter-professional challenges may arise between sustainability professionals and financial practitioners in the production of climate risk reporting, given different ontological views of climate risk materiality. Views of when climate risk is or is not material have direct bearing on disclosure and those views may vary between sustainability and financial practitioners. The prospect of workable collaborations between practitioners is a source of transition challenge reflected in Esbjorn-Hargens' multi-object phenomenon.

2.9   Ontological Standardization

Since the publication of the Task Force Report on Climate-Related Disclosures in 2017, corporate reporting groups have promulgated the need for companies to align themselves with the recommendations of the TCFD in anticipation of regulatory action. Where companies previously might have had the latitude of voluntarily and subjectively interpreting climate exposures, most likely by following the E (for Environment) under ESG guidelines, or in legacy corporate social responsibility reporting documents, new characterizations of climate risk and the processes by which climate risk management will be assessed by reporting audiences have emerged. At the time of writing public companies no longer have reporting flexibility over climate risk; instead, they are guided by precise definitions of what constitutes a climate risk and its treatment, i.e., through mitigation, and how precisely the company must report on it. These actions on the part of the global Financial Stability Board (FSB) both enliven and simplify desired approaches to align and systematize climate risk with risk practice, via a deliberate ontological 'standardization.' Standardizing meanings of risk in this respect is in stark contrast to constructionist views of climate risk, and in the process, eliminates any possibility of response optionality for the firm. Actions taken by the TCFD, purportedly in reaction to slow mitigation efforts, and for the need for financial transparency among investor audiences, have ushered in a new ontology of climate risk. This directly affects the cognition of corporate actors to standardize their risk management practices accommodating it. The transforming impact of this standardization is another example of how business is confronted by a new ontology of climate risks.

In further discussion, language use in climate risk narratives may have a similar influence on cognitive processes of organizations, discussed next.

2.10   Cognition
and Semantic-Linguistic Manipulations

Two useful examples in literature illustrating cognitive manipulation by language use involve (a) terminology use and (b) semantically non-equivalent and contradictory expressions of climate risk issues. In the first instance, current climate risk terminology involves terms such as carbon offsetting, social cost of capital, emissions abatement, and decarbonization pathways. Some if not all terms contribute to what some researchers call the climate narrative contained in the story telling frame of whichever climate risk issue is being discussed (by whichever group is doing the story telling).

In the second instance, climate language is filled with semantically non-equivalent, contradictory, and confusing expressions and terms that may misguide response actions of firms. For example, the reference to 'the dichotomy of adaptation versus mitigation' in Biesbroek et al. (2009) suggests response actions to reducing climate risks have a zero-sum quality eliminating multiple response strategies. A brief snapshot of the linguistic history of climate language illustrates more.

Language use of climate terminology was noted in literature in the last decade—the warmest decade recorded globally since 1880 (Schmunk, 2010). The term 'global warming' was popular and coined by Walter Broecker of Columbia University who commented: 'It is possible that we are on the brink of a several decade long period of rapid warming'

(Broecker, 1975, p. 462). The term climate change entered the lexicon of climate science in the 1990s when the IPCC formally stated that 'the side effects of global warming such as melting glaciers, heavier rainstorms and more frequent drought were part of an emerging future climate state' (Solomon et al., 2007).

A point to consider is that empirical work showed that 'global warming' and 'climate change' were not semantically equivalent terms, and furthermore produced different connotations eliciting different reactions in people (Bostrom et al., 1994). A popular trivialization of climate risk is found in the divide between partisans over the use of chosen climate frames related to how serious sounding the term is, i.e., climate change is viewed as more serious than 'global warming' (Boykoff & Boykoff, 2007; Villar & Krosnick, 2011).

Another illustration of language use that affects cognition is the 'future time reference of risk' in producing the effect of associating the future with the present. As previously discussed, definitions of risk are instrumental to how risk is perceived and enacted upon, through conscious human cognitive deliberations. Disassociating the future with the present—as the term climate change does—provides a language structure that makes the future farther away to the present state, and reduces what linguists call future time reference (FTR). Empirical work done by Chen et al. (2019) suggests that future-oriented behavior is fostered more effectively when terminology combines future and present state constructs, that is, to bring the future closer to the present.

Hence, in combination the words climate (the current) and risk (the future) bring the future closer to the present. This effect of language structure on this habit of speech can cause people to value future rewards and lead speakers to take more future-oriented actions (Chen et al., 2019). In practice, when a phenomenon is regarded as a business risk, the management of those risks involves anticipating the future, and preventing or reducing those risks before problems occur in the future, rather than responding and reacting to threats after the fact when the damage has already been done (Barton et al., 2002).

The discursive construction of climate change as a risk, provides a schema of interpretation that helps individuals to perceive and label it (Lefsrud & Meyer, 2012). Risk has long been the language and a managerial object of auditors, insurers, and risk managers; however, it is has become a new frame in climate policy and increasingly a part of business rhetoric (Kasperson et al., 2012).

As stated by the IPCC, climate risk is the unifying theme and the 'desired approach' (Intergovernmental Panel on Climate Change, 2014) reflecting how global governance groups now regard the future uncertainties associated with climate risk. Supranational groups such as the World Bank, go so far as to refer to climate risk as a 'force majeure' phenomena, exacting overwhelming force on entities beyond what can be comprehensively planned for (The World Bank, 2021).

While the question of who has definitional authority for climate risk terminology is still an open question, how groups choose to frame the characteristics of climate risk through language use have the potential to alter cognition through manipulation of semantic treatments.

2.11   COGNITION AND RISK BELIEF

Underlying cognitive processes are risk beliefs held to be true by individuals or groups (Sidortsov, 2014; Torcello, 2016.) Defined as the mental acceptance or conviction in the truth or actuality of some idea, risk beliefs are significant because they also provide the basis for individuals to understand the world and act within it (Bell et al., 2006; Gann & Matlock, 2014). Risk beliefs, or perhaps more realistically risk belief systems, provide the mental scaffolding for appraising the environment, explaining new observations, and constructing a shared meaning of the world. Given that cognition influences behavior, and those cognitions are driven by beliefs held to be true by individuals and have a causal effect on corporate performance (Goodhew, 1998), it is understandable why management cognition and risk beliefs are important constituents of risk-based decision-making. For example, the options a firm exercises to reduce climate risk exposures, will be at least based on management's perception of the risk at hand. Risk beliefs may include the closely held conviction by company actors that entirely or substantially new business resources, processes, and risk practices are needed by the firm to pursue low carbon initiatives, or that the acquisition and development of technical knowledge and expertise is needed to stress test system infrastructure under warming temperature scenarios (Dowbiggin, 2018). Risk beliefs may drive decision-making about what level of importance is to be placed on motivating Scope 3 supply chain partners to decarbonize and to improve corporate reporting on carbon performance (Damert et al., 2017). These examples imply cognitive cohesion or shared views of company objectives.

Where complications arise are in the circumstances of mixed risk beliefs and the lack of cognitive capacity on the part of actors, to discern a unified corporate approach to reducing climate risk. As articulated in TCFD risk guidance documents, successful climate risk management is likely to be best served by an 'all-in' cross company collaboration and interconnectedness to manage climate risks. This enterprise-wide view of cognition implies a requirement for cognitive fusion with unified risk beliefs, discussed next.

2.12   COGNITIVE FUSION
AND UNDERLYING RISK BELIEFS

Given that risk beliefs underpin cognition, companies working diligently to meet new climate risk management expectations, are better off understanding what risk belief systems are being operationalized within the organization. Understanding which corporate constituents hold which risk beliefs can be mapped by cognitive mapping tools to elicit mental models, defined as the 'internal, cognitive representation of a state of affairs in the external world' (Craik, 1943; Wood et al., 2012). Tools such as mental modeling to identify inconsistencies and conflicts produced by variations in perceptions and beliefs of company actors will help build cognitive unity and interconnectedness to address climate risks across the organization.

By far more compelling, is the challenge of complexity factors of climate risk on cognitions, discussed next.

2.13   COGNITIVE MASTERY OVER COMPLEXITY

The level of cognitive mastery, i.e., ability to understand and interpret complex problems and solutions, is critical to a rational, science-informed view of climate risk, to the selection of response actions for low carbon initiatives, and to the longer term management of a decarbonization strategy for the firm. Despite strategic intention, the realistic transformability of the organization to decarbonize may depend on how corporate actors think about any number of factors enabling or preventing the firm's carbon reduction efforts. This view of cognitive competency is supported by social cognitive theory especially when it exists in complex systems (Wood & Bandura, 1989). I posit that higher order management cognition is needed to manage the complexity factors brought about by risk exposures appearing on multiple fronts of the company. Cognitive competency levels will necessarily need to be improved to buttress and manage organizational system shocks. Organizational system shocks may occur due to new corporate strategies, organizational design features, e.g., new climate office, operational policies, and new management control practices.

Five examples of drivers producing cognitive transition may include: (1) entirely or substantially new risk beliefs; (2) elevation in the level of emphasis on group effort and risk belief homogeneity; (3) entirely or substantially new strategic objectives; (4) a level of emphasis on rapid action, and (5) entirely or substantially new internal risk practices. The extent and involvement of and impact on Scope 2 and 3 partners and the extent and involvement of and impact on customer and investor expectations are two further examples of drivers causing what is likely to be a cognitive transition at the firm level (Fig. 2.1).

Furthermore, strategic outcomes of the firm are doubly concerned with carbon reduction, precaution, and loss control in combination with profit and risk taking. This implies cognitive flexibility or an ambidexterity to accommodate paradoxical views of company objectives. Where prior gain–loss trade-offs might have existed in competitive advantage literature, decarbonization goals are now negotiated as a competing, paradoxical aim of the firm. Furthermore, unlike slower moving organizational transformations such as globalization and technology adoption, rapid

---

* Entirely or substantially new risk beliefs

* Level of emphasis on group effort and risk belief homogeneity

* Entirely or substantially new strategic objectives

* Level of emphasis on rapid action and temporal changes

* Entirely or substantially new internal risk practices

* Extent and involvement of and impact on Scope 3 supply chain partners

* Extent and involvement of and impact on customer and investor expectations

---

Fig. 2.1 Drivers of cognitive transitions regarding climate risk challenges, by author

deployment of new practices, approaches, and supporting routines are required. Public companies have disclosure deadlines, stringent reporting requirements supported by metrics and tools that are new to the enterprise. Medium and smaller sized privately held companies connected with the carbon chain of larger entities must adopt similar practices.

2.14   Conclusion

Multiple cognitive transitions are likely to occur in organizations as a result of climate risk exposures. Driven by complexity factors, risk culture changes, risk translations, risk role and ownership changes and the ontological standardization of climate risk, cognitions and risk beliefs are expected to materialize (Wiltshire et al., 2014).

References

Barton, T., Shenkir, W., & Walker, P. (2002). *Making enterprise risk management pay off: How leading companies implement risk management*. Prentice Hall.

Battiston, S., Dafermos, Y., & Monasterolo, I. (2021). Climate risks and financial stability. https://doi.org/10.1016/j.jfs.2021.100867

Beck, U. (1999). *World risk society*. Polity Press.

Bell, V., Halligan, P. W., & Ellis, H. D. (2006). Explaining delusions: A cognitive perspective. *Trends in Cognitive Sciences, 10*(5), 219–226.

Bevere, L., & Weigel, A. (2021). *Natural Catastrophes in 2020*. Swiss Re Institute. https://www.swissre.com/institute/research/sigma-research/sigma-2021-01.html

Biesbroek, G. R., Swart, R. J., & Van der Knaap, W. G. (2009). The mitigation–adaptation dichotomy and the role of spatial planning. *Habitat International, 33*(3), 230–237. https://doi.org/10.1016/j.habitatint.2008.10.001

Bostrom, A., Morgan, M. G., Fischhoff, B., & Read, D. (1994). What do people know about global climate change? 1. *Mental Models. Risk Analysis, 14*(6), 959–970.

Boykoff, M. T., & Boykoff, J. M. (2007). Climate change and journalistic norms: A case-study of US mass-media coverage. *Geoforum, 38*(6), 1190–1204. https://doi.org/10.1016/j.geoforum.2007.01.008

Broecker, W. S. (1975). Climate change: Are we on the brink of a pronounced global warming? *Science, 189*(4201), 460–463. https://doi.org/10.1126/science.189.4201.460

Chandler, A. D. (1962). *Strategy and structure: Chapters in the history of the industrial empire*. MIT Press.

Chen, J. I., He, T. S., & Riyanto, Y. E. (2019). The effect of language on economic behavior: Examining the causal link between future tense and time preference in the lab. *European Economic Review, 120,* 103307. https://doi. org/10.1016/j.euroecorev.2019.103307

Craik, K. J. W. (1943). *The nature of explanation.* Cambridge University Press.

Damert, M., Paul, A., & Baumgartner, R. J. (2017). Exploring the determinants and long-term performance outcomes of corporate carbon strategies. *Journal of Cleaner Production, 160,* 123–138.

Dowbiggin, A. (2018). *Climate risk perceptions in the Ontario (Canada) electricity sector* (Doctoral Dissertation). Heriot Watt University, Heriot Watt Digital Archive.

Esbjörn-Hargens, S. (2010). An ontology of climate change. *Journal of Integral Theory and Practice, 5*(1), 143–174.

Fischhoff, B., Watson, S. R., & Hope, C. (1984). Defining risk. *Policy Sciences, 17*(2), 123–139. https://doi.org/10.1007/BF00146924.S2CID189827147

Gann, T., & Matlock, T. (2014, July 23–26). The Semantics of Climate Change and Global Warming. In P. Bello, M. Guarini, M. McShane, & Brian Scassellati (Chairs), *Proceedings of the Annual Meeting of the Cognitive Science Society* [Symposium]. Cognitive Science Society 36th Annual Conference, Quebec City, Canada.

Gasbarro, F., & Pinkse, J. (2015). Corporate adaptation behaviour to deal with climate change: The influence of firm specific interpretations of physical climate impacts. *Corporate Social Responsibility and Environmental Management, 23*(3), 179–192. https://doi.org/10.1002/csr.1374

Goodhew, G. (1998). *Cognition and management: Managerial cognition and organisational performance* (Doctoral dissertation). University of Canterbury, University of Canterbury Research Repository.

Intergovernmental Panel on Climate Change. (2012). *Managing the Risks of Extreme Events and Disasters to Advance Climate Change Adaptation.* https://www.ipcc.ch/site/assets/uploads/2018/03/SREX_Full_Report-1.pdf

Intergovernmental Panel on Climate Change. (2014). *Climate Change 2014: Mitigation of Climate Change.* https://www.ipcc.ch/site/assets/uploads/2018/02/ipcc_wg3_ar5_full.pdf

Kahneman, D., & Tversky, A. (1979). Prospect theory: An analysis of decision under risk. *Econometrica, 47*(2), 263–292. https://doi.org/10.2307/1914185

Kaplan, S., & Garrick, B. J. (1981). On the quantitative definition of risk. *Risk Analysis, 1*(1), 11–27.

Kasperson, J. X., Kasperson, R. E., Pidgeon, N., & Slovic, P. (2012). The social amplification of risk: Assessing 15 years of research and theory. In *Social contours of risk* (pp. 217–245). Routledge.

Kelly, G. (1955). *Personal construct psychology*. Norton.

Lefsrud, L. M., & Meyer, R. E. (2012). Science or science fiction? Professionals' discursive construction of climate change. *Organization Studies, 33*(11), 1477–1506. https://doi.org/10.1177/0170840612463317

Markus, H., & Zajonc, R. B. (1985). The cognitive perspective in social psychology. *Handbook of Social Psychology, 1*(1), 137–230.

Möller, N., & Hansson, S. O. (2008). Principles of engineering safety: Risk and uncertainty reduction. *Reliability Engineering & System Safety, 93*(6), 798–805.

Outhwaite, W. (1987). *New philosophies of social science: Realism, hermeneutics and critical theory*. Macmillan Publishers.

Sandman, P. M. (1987). Risk communication: Facing public outrage. *Environmental Protection Journal, 13*(21), 21.

Schmunk, R. B. (2010, January 21). *2009: Second warmest year on record; End of warmest decade*. NASA. http://www.giss.nasa.gov/research/news/20100121/

Sharma, S. (2000). Managerial interpretations and organizational context as predictors of corporate choice of environmental strategy. *Academy of Management Journal, 43*(4), 681–697.

Sidortsov, R. (2014). Reinventing rules for environmental risk governance in the energy sector. *Energy Research & Social Science, 1*, 171–182. https://doi.org/10.1016/j.erss.2014.03.013

Simon, H. A. (1955). A behavioral model of rational choice. *The Quarterly Journal of Economics, 69*(1), 99–118. https://doi.org/10.2307/1884852

Slovic, P., & Peters, E. (2006). Risk perception and affect. *Current Directions in Psychological Science, 15*(6), 322–325. https://doi.org/10.1111/j.1467-8721.2006.00461.x

Smircich, L., & Stubbart, C. (1985). Strategic management in an enacted world. *Academy of Management Review, 10*(4), 724–736.

Solomon, S., Qin, D., Manning, M., Averyt, K., & Marquis, M. (2007). *Climate change 2007: The physical science basis: Working Group I Contribution to the Fourth Assessment Report of the IPCC*. Cambridge University Press.

Stubbart, C. I. (1989). Managerial cognition: A missing link in strategic management research. *Journal of Management Studies, 26*(4), 325–347. https://doi.org/10.1111/j.1467-6486.1989.tb00732.x

The World Bank. (2021). *The Force Majeure Checklist*. https://ppp.worldbank.org/public-private-partnership/library/force-majeure-checklist-and-sample-wording

Torcello, L. (2016). The ethics of belief, cognition, and climate change pseudoskepticism: Implications for public discourse. *Topics in Cognitive Science, 8*(1), 19–48.

Villar, A., & Krosnick, J. A. (2011). Global warming vs. climate change, taxes vs. prices: Does word choice matter? *Climatic Change, 105*(1), 1–12. https://doi.org/10.1007/s10584-010-9882-x

Wald, A. (1939). Contributions to the theory of statistical estimation and testing hypotheses. *The Annals of Mathematical Statistics, 10*(4), 299–326.

Weick, K. E. (1979). *The social psychology of organizing*. Addison-Wesley.

Weinhofer, G., & Busch, T. (2013). Corporate strategies for managing climate risks. *Business Strategy and the Environment, 22*(2), 121–144. https://doi.org/10.1002/bse.1744

Wiltshire, T. J., Neville, K. J., Lauth, M. R., Rinkinen, C., & Ramirez, L. F. (2014). Applications of cognitive transformation theory: Examining the role of sensemaking in the instruction of air traffic control students. *Journal of Cognitive Engineering and Decision Making, 8*(3), 219–247. https://doi.org/10.1177/1555343414532470

Wood, M., Bostrom, A., Bridges, T., & Linkov, I. (2012). Cognitive mapping tools: Review and Risk management needs. *Risk Analysis, 32*(8), 1333–1348. https://doi.org/10.1111/j.1539-6924.2011.01767.x

Wood, R., & Bandura, A. (1989). Impact of conceptions of ability on self-regulatory mechanisms and complex decision making. *Journal of Personality and Social Psychology, 56*(3), 407–415. https://doi.org/10.1037//0022-3514.56.3.407

Wood, R. E., Mento, A. J., & Locke, E. A. (1987). Task complexity as a moderator of goal effects: A meta-analysis. *Journal of Applied Psychology, 72*(3), 416–425. https://doi.org/10.1037/0021-9010.72.3.416

CHAPTER 3

Resource Challenges

Abstract Views of the resource theory landscape to illuminate the linkages between resources and corporate objectives are presented. Definitional conceptualizations about organizational resources are offered as is a summary view of resource-based theories. Resource-based theories focus on the internal factors within organizations, emphasizing the firm's own decisions and competencies rather than its environment. This includes the organizational resources entities leverage for performance improvements. Using recent sources of empirical work, corporate mitigative efforts suggest a possible misalignment with the original conceptualizations of RBV and its variants including the natural resource-based view.

Keywords Carbon management · Carbon accounting · Resource-based theory · Natural resource-based theory · Complementary assets · Dynamic capabilities · Climate risk

3.1 INTRODUCTION

In the prior chapter, discussions were extended on how the transition challenges of climate risk create transforming effects on management cognition and risk beliefs. Management cognition and the underlying risk beliefs among actors are likely to shift from prior ways of thinking

© The Author(s), under exclusive license to Springer Nature Switzerland AG 2021
A. Dowbiggin, *Climate Risk and Business*,
https://doi.org/10.1007/978-3-030-78244-3_3

given the organizational shock and the onset of organizational complexity factors involved in managing climate risk. Drivers for cognition shifts to the complex level of transformative change were presented, as were discussions about cognition linkages with manipulations found in epistemological, ontological, and semantic treatments of climate risk. Created by urgent and coercive reporting and carbon enactments by financial risk governance groups, i.e., the FSB, TCFD, an 'ontological standardization of climate risk' has emerged. Forced constructions of what climate risk is, has created a new ontology of climate risk which has distinct implications for management cognition and climate risk beliefs.

The introduction of new risk categories to an organization represents a major transition in organizational priorities and management practices. The challenges that climate risk presents to business will most likely be varied, vast, and complex. As proposed at the outset of this book, I proffer a fivefold thought experiment about how climate risk will theoretically challenge business. Reflecting on new climate risk mitigation and reporting requirements, one may contemplate what the consequences will be for public companies and the medium and smaller entities connected to carbon chains of larger companies.

3.2   Organization of This Chapter

In this chapter, discussions focus on organizational resources, assets, processes, and practice transitions at the firm level. I proffer those corporate responses classified as either reactive or proactive actions to reduce climate loss exposures will give rise to new and transitional adjustments to organizational resources.

The progression of this chapter begins with a view of the theoretical landscape to illuminate the linkages between resources and corporate objectives. Definitional conceptualizations about organizational resources are offered as is a summary view of resource-based theories. Resource-based theories focus on the internal factors within organizations, emphasizing the firm's own decisions and competencies rather than its environment (Hoskisson et al., 1999). This includes the organizational resources entities leverage for performance improvements.

Using recent sources of empirical work, corporate mitigative efforts suggest a possible misalignment with the original conceptualizations of RBV and its variants including the natural resource-based view, conceived by Hart (1995).

3.3 THEORETICAL LANDSCAPE

As corporate response actions to new climate risk expectations are unfolding, the reliable picture of how companies initiate or maintain internal mitigative resources is not well understood. The underreporting in literature and practitioner journals and the lack of detailed operational disclosure of firms with decarbonization programs, obliterates a clear view of current corporate mitigation resources and activity. This leaves the full extent of resource transitions currently underway moderately under-acknowledged and potentially misunderstood. Noted are studies where carbon reduction programs have not contributed to successful outcomes (Doda et al., 2016). Certainly, as more embedded field level studies are undertaken, a clearer picture of the linkages between resource theory and mitigation effort would be revealed. This would assist in questioning the transition challenges associated with resource theory explanations. This is not to diminish prior work but to contribute useful new insights of the extent to which RBV supports current mitigation efforts driven by resources. Theoretical constructions are vital to the study of business and climate risk challenges, given that the full extent of business response actions to climate risk is still unfolding. Furthermore, it is important to remember that theory is not conjecture but speculation rooted in the logic of prior work and brings with it an expected trajectory of theory improvement. Closing this theoretical gap would become an important source of academic enlivenment as more empirical evidence is analyzed and published in literature.

3.4 RESOURCE THEORIES

Resource theories expound on the intraorganizational drivers for resource deployments, and about how firms adjust internal resources and proce-dures to meet new requirements. They are rooted in a resource-centric view of organizational success and are visible in literatures of the resource-based view of the organization. Resource theories focus on the internal factors within an organization which contribute to organizational success, emphasizing the firm's own decisions and competencies rather than its environment (Hoskisson et al., 1999). To be clear on what is meant by resources explicated in theory, an explanatory view is offered next.

3.5   Resources as Assets, Processes, and Practices

Using definitional concepts offered by Arend and Levesque (2010), a resource is defined as an asset the firm possesses and manipulates for competitive advantage. Resources can be tangible such as physical property, or as an asset having transactional value. In the Arend and Levesque (2010) view, resources can also be intangible assets, such as employees' skills and expertise. Christmann (2000) asserts resources can be understood also as complementary assets, where their presence makes resources more valuable to the firm. Complementary assets are underscored by the perspective that assets alone do not ensure success but that the addition of differentiated assets does.

As well, resource theories conceptualize resources as processes explicated by Hart and Dowell (2011). Processes are understood as well-defined, repeatable steps that achieve a predictable result. Occurrences of resources as processes in current organizational environments are evident in literature.

For example, emission control processes involve systematic data collection and stepwise evaluative processes of carbon measurement. The lifecycle assessment processes of automotive facilities examined by Gebler et al. (2020), support mitigation objectives by examining all of the resource flows to decarbonize automotive production. Hart and Dowell (2011) illuminate how product ideation processes can assist green product innovation and the procurement of low carbon raw materials. Processes for the measurement of GhG are problematized by Pascaris and Pearce (2020) with respect to competing methodologies of 'point source methods' versus 'bottleneck methods' of calculating emissions output of polluting entities.

In further distinction, resource theories involve organizational practices that can be inextricably linked to processes and assets to form resource domains, as conceptualized by Buysse and Verbeke (2003). Resource domains are bounded spaces containing bundled resources that operationalize features of corporate strategy (Buysse & Verbeke, 2003). A case in point is Bui and Villiers' (2017) survey of 30 New Zealand firms using carbon management control systems for multiple boundaries and performance-setting objectives in strategy and operations controls. The survey revealed variations in the level of integration of control systems according to corporate characteristics of size and carbon intensity;

high carbon intensity firms showed the greatest strategic and operational integration of control systems—suggesting that carbon management resources and processes were viewed as a core business activity. Backman's (2017) construction of the GISTe model (representing governance, information management, systems, and technology) illustrates the idea of a resource domain where multiple assets, processes, and practices are operationalized in a bundled format. Backman et al. (2017) conceptualized the GISTe model to analyze mitigation strategies of 552 firms participating in the Carbon Disclosure Project (CDP). The model contextualizes which organizational resource domains are viewed as being critical for mitigation using the real-world empirical data set of the CDP. This example identifies and classifies resource domains that operationalize carbon reduction strategies according to self-reported company registrants and illustrates how combined interdependencies of resource assets, processes, and practices work together in mitigation effort. In another example, Hahn et al. (2015) deconstruct the resource domain of carbon accounting with a combinative array of assets, processes, and practices, in the following manner: In carbon accounting, the technological and financial management control and information systems are the tangible assets, the systematic collection of data to measure direct and indirect carbon emissions of the firm is the process, and the actions of measuring and processing carbon-related information are the practice subsumed under carbon accounting.

Conceptually, the configuration of resources as bundled resource arrangements in domains raises questions for further research. Which resource domains, or specific spheres of resources and activities, are instrumental in supporting carbon management within firms? What resource elements are logically grouped together in organizational spheres, and to what extent can the resource domain construct be abstracted for company actors whose tasks and accountabilities are situated simultaneously in several domains?

3.6   Resource-Based View

The resource-based view (RBV) of the firm is an enduring perspective on organizational success conceptualized by Penrose in 1959 and elaborated on by 60 years of additional analysis and commentary by theorists including Jay B. Barney, George S. Day, Gary Hamel, Shelby D. Hunt, G. Hooley, and C. K. Prahalad.

Its endurance as a dominant resource theory continues to be supported (and critiqued) to such an extent that RBV scholars have produced dozens of theoretical variations and application domain studies that isolate valuable sources of enabling characteristics of successful firms (Esteve-Pérez & Mañez-Castillejo, 2008). Some of these include core competencies (Prahalad & Hamel, 1990), strategic factor markets (Barney, 1986), corporate climate (Hansen & Wernerfelt, 1989), and intangible assets. Contributions to RBV over the last 20 years have emphasized CEOs as a firm resource (Castanias & Helfat, 1991), organizational identity as a core competency (Fiol, 1991), the beneficial outcomes of combinative capabilities (Kogut & Zander, 1992), and the performance outcomes of merging RBV with institutional theory (Oliver, 1997). Additional insights were shown by applying RBV to practitioner processes and routines such as human resources (Wright et al., 2001), strategic entrepreneurship, and procurement (Makadok, 2001).

3.7   New Paradigm

It seems reasonably clear that RBV's endurance as a theory is at least due to its explanatory power to account for multiple applications in practice, and its intuitive simplicity and appeal for understanding corporate success in the competitive advantage paradigm of the globalization decade in the 90s and the early twenty-first century (Bowman & Ambrosini, 2003). In today's world, new definitions of corporate success are emerging to include climate risk mitigation efforts in a duality of purpose for business.

Corporate objectives in the current climate risk context are those which foster transitions to a low carbon economy and involve management resources which assist in mitigation efforts, e.g., carbon management programs (de Abreu et al., 2017), carbon emissions management control systems (Bui & Villiers, 2017), and specialized management control systems (Kumarasiri & Gunasekarage, 2017).

Does this suggest a paradigmatic shift of compelling new accommodations in practice? Accounting for carbon management as a process initiative for decarbonization strategy underscored by cautionary approaches and loss control in contrast to or in combination with growth, profit, and risk-taking, suggests new approaches to the objective setting are visible in practice. Is RBV still theoretically useful as an explanatory force for organizational response to both climate threat exposures alongside of market maximization strategies?

3.8   RESOURCE CRITERIA

Another unresolved issue involves resource criteria specified in RBV and presents a new question: Would the RBV criteria for company differentiation via valuable, rare, imperfectly imitable, and non-substitutable resources of the firm ('VRIN'), remain unchanged in view of potentially shared decarbonization approaches, e.g., pooled carbon technologies and if so, how? Companies that 'don't go it alone' in decarbonization, may develop firm-level capacity in coordination with sector partners and supply chain groups (Dowbiggin, 2018). In those circumstances when interorganizational linkages emerge and critical interdependencies develop in the institutional landscape, will it render VRIN criteria impractical, or possibly obsolete?

3.9   UNDER-RESOURCED EFFECTS

A 2021 snapshot of Carbon Disclosure Project report of companies raises other questions about RBV and mitigation effort and reporting. At the time of CDP publication, 4,500 global reporting companies made their GhG emissions data available on the voluntary CDP registry for public analysis. There, corporate response on corporate mitigation implementation is lower than expected (Anderson, 2019). Responding to the upcoming 2025 accountability deadline for full financially standardized climate risk disclosures might have suggested a higher performance in reporting. These are surprising findings.

While low emissions reporting levels imply slow carbon management reporting response and may even cause one to speculate on lower performance among non-reporting corporates, there may be any number of explanations. Companies may be stuck in a messy early life cycle stage of carbon strategy formulation, a decision-making impasse about priority issues, poor coordination of efforts, and deliberations over launching process initiatives that support emissions reporting. It may be that net zero transition roadmaps are absent, underfunded, or simply not coordinated or leveraged. Reluctance to disclose GhG emissions data may be the outcome of many factors occurring inside or outside of the firm. The question, as it would be for RBV scholars to ask, is what resource domain(s) within companies are supporting climate mitigation efforts, and what intraorganizational factors are constraining those efforts? (Kerber & Buono, 2004).

3.10   Divergence in Resource Theories

Resource theory (RBV) as first conceptualized by Penrose (1959) theorizes that corporate success is influenced by how intraorganizational resources are coordinated and managed within firms, rather than by how external factors exert pressures on business, as described in institutional theory. Resource theories embody organizational strengths and capabilities as the means by which firms achieve their own objectives. Implied in Penrose (1959), resource theory promotes a view of organizational life invariant to external events or changing conditions in the business environment. Hart (1995) challenged the assumption that corporate successes were achieved at the implied exclusion of outside influences, producing a major divergence in resource-based theory. His ideas conceptualized in the Natural Resource-Based View (NRBV) raised the importance of the firm's interaction with its natural environment, at a time of growing concern for sustainability issues. At the time, Hart states, 'It is likely that strategy and competitive advantage in the coming years will be rooted in capabilities that facilitate environmentally sustainable economic activity – a natural-resource based view of the firm' (Hart, 1995, p. 991).

3.11   Dynamic Capabilities, Complementary Assets

Other spin-off perspectives yielding to NRBV notably the Dynamic Capabilities view (Eisenhardt & Martin, 2000; Teece et al., 1997; Winter, 2003) and Complementary Assets (Christmann, 2000) suggested strategic capabilities of the firm are activated by different processes and activities related to environmental resource management in the sense that the degree to which resources add value, may depend on the presence of complementary assets and the dynamic capabilities of the organization. Complementary assets are understood as those acquired and utilized to support differentiated sustainability objectives of corporations. In other words, a firm's environmental management focus on identifying best practices through the use of complementary assets, can potentially and simultaneously reduce the negative impact of the firm's activities on the natural environment and concomitantly contribute to competitive advantage. In particular, three strategic capabilities of firms, befitting the definition of complementary assets and dynamic capabilities were proffered by Hart and Dowell (2011) as exemplary for improved firm-environmental interactions, as follows: (1) pollution prevention—designed to reduce costs; (2) resource acquisitions of 'green' raw materials, intended to spark consumer appeal; and (3) life cycle assessment programs, to produce differentiated product designs (Hart & Dowell, 2011). Hart and Dowell (2011) viewed them as differentiating processes and activities for competitive advantage, at a time in management history that focused on business impacts on the environment.

3.12   THEORY REFURBISHMENT?

Under current conditions of climate risk management and expectations rooted in not only inside-out views of business impacts embodied in ESG management and reporting, but also of outside-in views of climate impacts on future business, the NRBV may well be in need of refurbishment. Core concepts in RBV and NRBV are locked into impacts-out views of business activity which is incongruent with climate risk management. Interconnected perspectives, e.g., dynamic capabilities and complementary assets, need theoretical adjustments to accommodate dual-purpose objectives of carbon reduction and corporate performance. Reforms of resource theory and its variations to date could result in improved alignments with multiple corporate purposes. Extensions with organizational change theory may be required in a hybridized 'resource change' orientation. Climate risk brings elevated levels of complexity to firm environments manifested by increasing interactions and relationships between resources and processes, changing objectives, a broadening scope of accountability to account for carbon in supply chains, disclosure accountability to account for 'real' mitigation efforts that structurally adjust GhG emissions and in product innovations in the market.

3.13   NEW REQUIREMENTS

With climate risk, there are new requirements to accommodate competing corporate objectives rooted in dual paradigms of competitive advantage and risk management. On one hand, corporate objectives of growth, opportunity, and risk-taking rooted in competitive advantage now moderate and are moderated by objectives of risk control, loss prevention, and caution. This suggests an emergent duality of corporate purpose, with multiple trade-off tensions and a paradoxical view of corporate existence. Resources used to accommodate paradoxical purposes can better help us understand the enduring value of resource theories when modified for the climate emergency.

According to Kerber and Buono (2004), transitional challenges facing business are moderated by the internal infrastructure of resources that facilitate change and whether there are sufficiently appropriate resources available to support change. Kerber and Buono (2004) take into account how a company's internal infrastructure that facilitates change and the appropriate resources available to support change might respond to narrower schemes of risk—but not of complex set of climate risks facing business at this point in time. This might suggest that Kerber and Buono's (2004) work and other resource-based theories and theoretical spin-offs published prior to Paris Agreement expectations, face an inevitable discount, given the extent of new and evolving complexities metered in organizational life under climate risk.

Furthermore, when corporations underperform, as in the case of low emissions reporting, Beer (2001) suggests organizations are challenged to build new capabilities to enhance overall corporate capacity to do so. Building mitigative capacity is an implied objective among CDP reporting companies, and as such, involves managing resources and change in ways that are appropriate to carbon reduction goals.

Implications for resource theory with paradox are discussed next.

3.14 Resource Theory and Paradox

Prior to the onset of paradox literature conceptualized and credited mainly to M.W. Lewis (Lewis, 2000, 2017; Lewis & Grimes, 1999; Lewis & Smith 2014; Lewis et al., 2002) organizational tensions were treated as organizational problems needing resolution. Scholars assumed that optimal solutions could be defined by analyzing the contingencies to the problem. In contrast, paradox suggests that tensions persistently defy a solution, implying that paradox endures as an ongoing tension in organizational life (Smith & Lewis, 2011). Conceptual clarity was developed over time, clarifying that paradox can be distinguished from other related concepts such as dilemmas, i.e., a trade-off decision made between beneficial alternatives. Schad and Miron-Spektor (2020), who analyzed Lewis' lifetime research contributions to paradox theory, suggest trade-off decisions offer temporary reprieve while paradox is a special breed of tension that 'persists as a contradiction between interdependent elements' (Schad et al., 2016).

The paradoxical nature of competing corporate objectives that climate risk introduces, invites questions about how resource theories can account for them. Managing paradox as Miron-Spektor et al. (2018) suggests, can be done through framing, or a mental template that encourages recognition and embracing of contradictions, which helps actors replace fixedness and rigidity with flexibility and creativity (Miron-Spektor et al., 2011; Smith & Tushman, 2005). In other work, organizational ambidexterity—to handle both incremental and discontinuous innovation and change, is seen as a tool to manage paradox, at the organizational level (Tushman & O'Reilly, 1996).

Climate risk creates paradox in organizations in three ways. First the double materiality of the climate risk model, introduced in Chapter 1, exhibits paradoxical qualities as an ongoing tension between inside-out risks to the global environment, and an ongoing tension of the outside-in risks of physical climate impacts on organizational life. There are no trade-offs to be made between holistic sources of risk, exemplifying what Lewis refers to as a persistent tension between two interrelated elements.

Second, paradox is found in prior work where authors suggest mitigation and adaption responses are false dichotomies, suggested both cannot be sufficiently addressed at the same time, due to finite resources of the firm. Climate risk responses require mitigation and adaptation and are mutually important responses to climate risk. Non-binary approaches to the mitigation-adaptation trade-off can be discounted in favor of paradox.

Third, paradox is manifested in the duality of purpose to accommodate growth and risk-taking, in tension with corporate objectives of climate risk management and caution. Managing competing sets of corporate objectives involves changes to the organizational arrangement of those resources into supporting resource domains.

3.15   Conclusion

Companies have unprecedented resource and process requirements to utilize in climate risk mitigation efforts. Climate risk exposure is a new application domain for resource theories and as such, it challenges prior explanations. The relevant and concluding open question then becomes: What amount of relevance exists for resource theories conceptualized in the 80s and 90s, even with recent additive perspectives, to sufficiently account for the multiple organizational transitions underway in business?

References

Anderson, N. (2019). *IFRS Standards and climate-related disclosures*. International Financial Reporting Standards. https://www.pwc.ch/en/publications/2021/in-brief-climate-change-nick-anderson.pdf.

Arend, R. J., & Lévesque, M. (2010). Is the resource-based view a practical organizational theory? *Organization Science, 21*(4), 913–930. https://doi.org/10.1287/orsc.1090.0484

Backman, C. A., Verbeke, A., & Schulz, R. A. (2017). The drivers of corporate climate change strategies and public policy: A new resource-based view perspective. *Business & Society, 56*(4), 545–575. https://doi.org/10.1177/0007650315578450

Barney, J. B. (1986). Strategic factor markets: Expectations, luck, and business strategy. *Management Science, 32*(10), 1231–1241. https://doi.org/10.1287/mnsc.32.10.1231

Beer, M. (2001). How to develop an organization capable of sustained high performance: Embrace the drive for results-capability development paradox. *Organizational Dynamics, 29*(4), 233–247. https://doi.org/10.1016/S0090-2616(01)00030-4

Bowman, C., & Ambrosini, V. (2003). How the resource-based and the dynamic capability views of the firm inform corporate-level strategy. *British Journal of Management, 14*(4), 289–303. https://doi.org/10.1111/j.1467-8551.2003.00380.x

Bui, B., & De Villiers, C. (2017). Carbon emissions management control systems: Field study evidence. *Journal of Cleaner Production, 166*, 1283–1294.

Buysse, K., & Verbeke, A. (2003). Proactive environmental strategies: A stakeholder management perspective. *Strategic Management Journal, 24*(5), 453–470. https://doi.org/10.1002/smj.299

Castanias, R. P., & Helfat, C. E. (1991). Managerial resources and rents. *Journal of Management, 17*(1), 155–171. https://doi.org/10.1177/014920639101700110

Christmann, P. (2000). Effects of "best practices" of environmental management on cost advantage: The role of complementary assets. *Academy of Management Journal, 43*(4), 663–680. https://doi.org/10.5465/1556360

de Abreu, M. C. S., de Freitas, A. R. P., & Rebouças, S. M. D. P. (2017). Conceptual model for corporate climate change strategy development: Empirical evidence from the energy sector. *Journal of Cleaner Production, 165*, 382–392.

Doda, B., Gennaioli, C., Gouldson, A., Grover, D., & Sullivan, R. (2016). Are corporate carbon management practices reducing corporate carbon emissions? *Corporate Social Responsibility and Environmental Management, 23*(5), 257–270.

Dowbiggin, A. (2018). *Climate risk perceptions in the Ontario (Canada) electricity sector* (Doctoral Dissertation). Heriot Watt University.

Eisenhardt, K. M., & Martin, J. A. (2000). Dynamic capabilities: What are they? *Strategic Management Journal, 21*(10–11), 1105–1121. https://doi.org/10.1002/1097-0266(200010/11)21:10/11%3c1105::AID-SMJ133%3e3.0.CO;2-E

Esteve-Pérez, S., & Mañez-Castillejo, J. A. (2008). The resource-based theory of the firm and firm survival. *Small Business Economics, 30*(3), 231–249.

Fiol, C. M. (1991). Managing culture as a competitive resource: An identity-based view of sustainable competitive advantage. *Journal of Management, 17*(1), 191–211. https://doi.org/10.1177/014920639101700112

Gebler, M., Cerdas, J. F., Thiede, S., & Herrmann, C. (2020). Life cycle assessment of an automotive factory: Identifying challenges for the decarbonization of automotive production—A case study. *Journal of Cleaner Production, 270*, 122330.

Hahn, T., Pinkse, J., Preuss, L., & Figge, F. (2015). Tensions in corporate sustainability: Towards an integrative framework. *Journal of Business Ethics, 127*(2), 297–316.

Hansen, G. S., & Wernerfelt, B. (1989). Determinants of firm performance: The relative importance of economic and organizational factors. *Strategic Management Journal, 10*(5), 399–411. https://doi.org/10.1002/smj.4250100502

Hart, S. L. (1995). A natural-resource-based view of the firm. *Academy of Management Review, 20*(4), 986–1014. https://doi.org/10.5465/amr.1995.9512280033

Hart, S. L., & Dowell, G. (2011). Invited editorial: A natural-resource-based view of the firm: Fifteen years after. *Journal of Management, 37*(5), 1464–1479. https://doi.org/10.1177/0149206310390219

Hoskisson, R. E., Wan, W. P., Yiu, D., & Hitt, M. A. (1999). Theory and research in strategic management: Swings of a pendulum. *Journal of Management, 25*(3), 417–456. https://doi.org/10.1016/S0149-2063(99)00008-2

Kerber, K. W., & Buono, A. F. (2004). Leadership challenges in global virtual teams: Lessons from the field. *SAM Advanced Management Journal, 69*(4), 4.

Kogut, B., & Zander, U. (1992). Knowledge of the firm, combinative capabilities, and the replication of technology. *Organization Science, 3*(3), 383–397. https://doi.org/10.1287/orsc.3.3.383

Kumarasiri, J., & Gunasekarage, A. (2017). Risk regulation, community pressure and the use of management accounting in managing climate change risk: Australian evidence. *The British Accounting Review, 49*(1), 25–38.

Lewis, M. W. (2000). Exploring paradox: Toward a more comprehensive guide. *Academy of Management Review, 25*(4), 760–776. https://doi.org/10.5465/amr.2000.3707712

Lewis, M. W. (2017, October 15). *Paradoxes of business schools.* Global Focus. https://www.globalfocusmagazine.com/paradoxes-business-schools/

Lewis, M. W., & Grimes, A. I. (1999). Meta triangulation: Building theory from multiple paradigms. *Academy of Management Review, 24*(4), 672–690. https://doi.org/10.5465/amr.1999.2553247

Lewis, M. W., & Smith, W. K. (2014). Paradox as a metatheoretical perspective: Sharpening the focus and widening the scope. *The Journal of Applied Behavioral Science, 50*(2), 127–149. https://doi.org/10.1177/0021886314522322

Lewis, M. W., Welsh, M. A., Dehler, G. E., & Green, S. G. (2002). Product development tensions: Exploring contrasting styles of project management. *Academy of Management Journal, 45*(3), 546–564. https://doi.org/10.5465/3069380

Makadok, R. (2001). Toward a synthesis of the resource-based and dynamic-capability views of rent creation. *Strategic Management Journal, 22*(5), 387–401.

Miron-Spektor, E., Gino, F., & Argote, L. (2011). Paradoxical frames and creative sparks: Enhancing individual creativity through conflict and integration. *Organizational Behavior and Human Decision Processes, 116*(2), 229–240. https://doi.org/10.1016/j.obhdp.2011.03.006

Miron-Spektor, E., Ingram, A., Keller, J., Smith, W. K., & Lewis, M. W. (2018). Micro foundations of organizational paradox: The problem is how we think about the problem. *Academy of Management Journal, 61*(1), 26–45. https://doi.org/10.5465/amj.2016.0594

Oliver, C. (1997). Sustainable competitive advantage: Combining institutional and resource-based views. *Strategic Management Journal, 18*(9), 697–713. https://doi.org/10.1002/(SICI)1097-0266(199710)18:9%3c697::AID-SMJ909%3e3.0.CO;2-C

Pascaris, A. S., & Pearce, J. M. (2020). US Greenhouse gas emission bottlenecks: Prioritization of Targets for climate liability. *Energies, 13*(15), 3932.

Penrose, E. T. (1959). *The theory of growth of the firm.* Blackwell Publishers.

Prahalad, C. K., & Hamel, G. (1990). The core competence of corporation. *The Harvard Business Review, 33*, 79–91.

Schad, J., Lewis, M. W., Raisch, S., & Smith, W. K. (2016). Paradox research in management science: Looking back to move forward. *Academy of Management Annals, 10*(1), 5–64. https://doi.org/10.5465/19416520.2016.1162422

Schad, J., & Miron-Spektor, E. (2020). Marianne W. Lewis: Paradoxes of change and changing through paradox. In D. B. Szabla (Ed.), *Palgrave Handbook of Organizational Change Thinkers* (2nd ed.). Palgrave Macmillan.

Smith, W. K., & Lewis, M. W. (2011). Toward a theory of paradox: A dynamic equilibrium model of organizing. *Academy of management Review, 36*(2), 381–403.

Smith, W. K., & Tushman, M. L. (2005). Managing strategic contradictions: A top management model for managing innovation streams. *Organization Science, 16*(5), 522–536.

Teece, D. J., Pisano, G., & Shuen, A. (1997). Dynamic capabilities and strategic management. *Strategic Management Journal, 18*(7), 509–533. https://doi. org/10.1002/(SICI)1097-0266(199708)18:7%3c509::AID-SMJ882%3e3.0. CO;2-Z

Tushman, M. L., & O'Reilly, C. A., III. (1996). Ambidextrous organizations: Managing evolutionary and revolutionary change. *California Management Review, 38*(4), 8–29. https://doi.org/10.2307/41165852

Winter, S. G. (2003). Understanding dynamic capabilities. *Strategic Management Journal, 24*(10), 991–995. https://doi.org/10.1002/smj.318

Wright, P. M., Dunford, B. B., & Snell, S. A. (2001). Human resources and the resource-based view of the firm. *Journal of Management, 27*(6), 701–721. https://doi.org/10.1177/014920630102700607

CHAPTER 4

Informal Risk Systems Challenges

Abstract Elevated expectations for how businesses will identify, assess, and manage climate-related risks as well incorporate climate risks into existing risk practices, are components of the emerging climate risk regime. Empirical and theoretical speculation in literature over risk culture changes, risk translations of climate risk, and contentions arising over risk roles and ownership constitute the combined focus of the discussions in this chapter.

Keywords Risk culture · Risk culture clash · Risk translation · Risk roles

4.1 Introduction

In the prior chapter, views of the resource theory landscape to illuminate the linkages between resources and corporate objectives, including decarbonization, were presented. Resource-based theories focus on the internal factors within organizations, emphasizing the firm's own decisions and competencies rather than its environment. This includes the organizational resources entities leverage for performance improvements and theoretically for corporate mitigative effort as well.

© The Author(s), under exclusive license to Springer Nature Switzerland AG 2021
A. Dowbiggin, *Climate Risk and Business*,
https://doi.org/10.1007/978-3-030-78244-3_4

4.2   Organization of This Chapter

The introduction of new risk categories to an organization represents a major transition in organizational priorities and management practices. The challenges of climate risk to business will most likely be varied, vast, and complex. As proposed at the outset of this book, I proffer a fivefold proposition about the ways in which climate risk will challenge business. Reflecting on the probable new organizational logics of corporate climate mitigation, one may contemplate what the consequences will be for business. This is the thought experiment of this book. So far, tentative propositions about potential cognitive and organizational resources have been discussed. A third transition challenge discussed in the next two chapters, that being of the impacts climate risk presents to firm-level risk management systems. Already noted, is the major challenge presented by climate risk in that it is of a fundamentally different nature to many other risks routinely priced by markets and involve considerably more complex underlying factors.

Elevated expectations for how businesses will identify, assess, and manage climate-related risks as well incorporate climate risks into existing risk practices, are components of the emerging climate risk reporting regime. Risk decisions based on internal risk management assessments form the basis for corporate climate strategy and affect risk disclosure. Disclosures directly affect public companies but have important consequences for medium and smaller non-reporting enterprises which are linked to reporting companies. Meeting institutional expectations will involve new and improved risk process initiatives that go beyond current business practices for all companies.

Accordingly, I suggest more specifically that the challenge of climate risk creates transitional pressures on both informal risk systems and formal risk systems of the organization. Given the probable scope and magnitude of how climate risk will alter company-wide risk management systems, I view a clear need to distinguish risk management as being both an informal and formal system. Perspectives on this distinction are inspired from literatures for risk practitioners, accounting, and management literature. Practical insights into the problems of resolving contentions in risk culture shifts, the equivocality of novel risks, and the arising tensions from interprofessional involvements in climate risk analysis and reporting are imported from and supported by divergent threads in literature streams.

This chapter addresses informal risk systems, understood as the differentiated and less tangible systems that exist outside the formal boundaries of risk management practices and structures, discussed next.

4.3 Informal Risk Systems

The mere mention of risk management systems to some may conjure up rigid and 'quantitatively enthusiastic' processes (Mikes, 2008). They may also involve supporting routines that 'inoculate the subjectivity' out of risk management methods (Dowbiggin, 2018). Yet under conditions of high complexity and rapid evolving change, organizational shock brought on by climate risk may have profound effects on risk culture.

Literature shows that risk culture involves idiosyncratic interactions and pragmatic relationship building done by risk actors under periods of organizational shock. Activities not directly involving conventional risk workstreams occur outside of formal risk management practices and can directly lead to changes in risk culture (Palermo et al., 2017). In other words, risk actor interactions colorize and are colorized by the risk culture of the firm, creating changes in organic processes resulting in a readjustment of risk culture under complex situations. Additionally, where new risks are not well understood, company actors may involve themselves in acts of risk translation in order to standardize approaches to interpreting risk events and their consequences. Furthermore, contentions among risk actors over risk roles and ownership due to collaborative constructions of risk materiality, may arise under climate risk circumstances.

Empirical and theoretical speculations in literature over risk culture changes, process transitions, risk translations of climate risk objects, and contentions arising over risk roles and ownership, constitute the combined focus of the discussions in this chapter.

4.4 Risk Specialists and Company Actors

My tentative proposition is that climate risk will alter the informal systems of risk culture at the firm level. The integration of climate risk into organizational risk management will involve the creation of a new risk culture and that the mechanics of building that culture will involve considerations similarly found in empirical work on risk culture reform. Construction of risk culture is likely to involve building a shared understanding of climate change concepts and risks across the company. Within firms, conventional risk management activities are usually conducted by specialist practitioners, i.e., risk actors, such as risk analysts, risk and compliance officers, internal audit, and corporate and external legal counsel, with reporting linkages to senior management, risk committees for oversight (Beasley et al., 2006; Subramaniam et al., 2015). However, given the broad implications of climate risk to affect multiple groups within the organizational structure, the identification, assessment, and management of climate risks will require involvement from various functions and departments within the company.

4.5   Organizational Culture

Organizational culture is conceptualized in literature as the pattern of shared values, beliefs, and assumptions considered to be the appropriate way to think and act within an organization. Prior definitional ideas about organizational culture has viewed it as a 'reified set of unified components pursuing shared goals' (Silverman, 1970), as 'dynamic systems of knowledge' (Smircich, 1983), or as 'interpretation systems whose social order must be negotiated' (Daft & Weick, 1984; Day & Day, 1977). Smircich (1983) referred to culture as 'reflexively defined by the interactions of actors in interpersonal episodes,' but Rose (1988) countered that coordination of interactions of actors is viewed as potentially problematic and is not assumed (Rose, 1988).

Increasingly in literature, conceptualizations have evolved to suggest that organizations are not sustained by one dominant unitary culture but are made up of multiple potentially conflicting cultures (Gregory, 1983; Martin & Siehl, 1983; Rose, 1988). Furthermore, where the nature of various multicultural arrangements of dominant and subcultures may exist in firms, an 'uneasy symbiosis' may result from tensions and conflicts produced among functions, e.g., in a practical example, tensions between the risk function versus business and corporate development functions (Martin & Siehl, 1983).

4.6   Risk Culture Transformation

My tentative proposition is that novel risks confronting business produce change in organizational risk culture that go beyond individual cognitions and belief systems. As discussed in Chapter 2, the impacts of climate risk on management cognition and risk beliefs held by individuals and group actors in complex settings, produce cognitive transformations based on changing perception and risk beliefs.

In this approach however, the scope is both broadened and narrowed to accommodate a larger cross-organizational view of a more specialized risk culture. Risk culture in organizations is understood as the set of encouraged and accepted behaviors, decisions, and attitudes taken toward the management of managing risk by an entity. Edgar Schein (2010) characterizes risk culture as being underpinned by the 'deeper layer of basic assumptions and beliefs' that are shared by members of an organization that operate unconsciously and that are defined in a taken for granted manner of an organization's self-conception. The study of risk culture and of risk culture reform brought about by increasing institutional complexity and far-reaching risk regimes, is thinly understood and almost absent in climate literature. Several reasons may explain this: (1) observation and measurement of risk culture transformation in rapidly evolving environments is inherently difficult to research as it may involve embeddedness and a willingness of the part of studied groups to cooperate in survey and data collection; (2) change in risk culture evolves over time, requiring costly longitudinal research approaches; and (3) risk culture changes are historically associated with high impact crises and force majeure-like phenomena, of which there have been relatively few studied in literature in recent history.

4.7 Forced Culture Change

However, some research reveals tentative approaches of how actors cope with forced requirements to reform risk practices under complex situations, that bear noting. One such example is the study of risk culture reform in financial groups in the aftermath of the 2008 financial crisis. The longitudinal study conducted by Palermo et al. (2017) showed how risk actors construct 'acts of intervention' when corporate objectives and processes uncouple in everyday practices under complex circumstances. When central bank regulation in the United Kingdom became a source of 'compliance-inducement' via the creation of standardized rules and practices under the Basel 3 accord, Palermo et al. (2017) explored how a reform of risk culture was instrumentalized by financial sector actors. Increasing institutional complexity of shifting corporate objectives was brought to bear on financial actors, instructed to 'clean up their culture' and reform their risk practices despite having 'an opaque relationship to outcome.'

A similar view of complexity and risk culture can be drawn between Palermo et al.'s (2017) UK banking crisis study and the expected mitigation effort of firms to execute and report on carbon reduction. Conflicting prescriptions of risk reduction and corporate risk-taking were shown to produce an ends-means uncoupling in risk culture, where actor practices de-couple or contradict corporate objectives of business growth. This end-means uncoupling transforms risk cultures and creates a potential gap between practices and organizational goals (Bromley & Powell, 2012; Palermo et al., 2017).

4.8   Patterns of Risk Culture Change

Three interpretations of Palermo et al.'s (2017) study, are potentially relevant to risk culture transitions under climate risk pressure.

One, the process of ends-means uncoupling by risk actors was patterned along with activities that evolved from an initial phase of idiosyncratic and organic logic, to one more systematized, engineered, and metric based. In the study where financial sector actors were encouraged to reform their risk culture to reduce organizational behaviors and decisions contributing to critical breakdowns in risk evaluations, the process started with organic activities such as risk conversations, pragmatic reaching out to colleagues, the acceptance of ambiguity in the objectives of the firm, a 'connecting of the dots' sense-making process, and the leveraging of existing resources and personal risk experiences.

Over time, an engineered phase emerged where risk culture processes transformed into more mechanistic activities of measuring, surveying, and incentive creation. This included a palpable search for analytical definitions of risk, coupled with a normative orientation of risk practitioners to produce diagnostic practices and a reductive theorization of risks.

The life cycle phasing of risk culture transition that Palermo et al. (2017) observed, may suggest a similar transition in climate-related risk culture by organizational actors grappling with conflicting prescriptions. The progression along such an uneven trajectory, deemed 'the messy initial phase in the lifecycle of a new concept' (p. 177), into one that is more reductive and diagnostic in orientation, invites theoretical speculation that firms in the early phase of climate risk assessment may reform extant risk cultures to meet new requirements.

The adjustment of organizational actors attuning to increased tensions of climate risk workstreams, perceived contradictions in company objectives, and coordination with specialists and generalists has implications for climate risk decision-making and the apparent time it may take for actors to mesh with new requirements.

4.9   Risk Culture as Intervention

Second, Palermo et al.'s (2017) view of institutional complexity offers a noteworthy parallel to the challenges facing business due to climate risk. Comparatively speaking, organizational complexity in the Palermo study was an outcome of shifting prescriptions to improve and adopt new risk practices. I proffer a similar argument that in the case of climate risk, new risk practices will be adopted, and that risk culture will evolve as an object of intervention to balance risk control and risk-taking. Indeed, as Palermo et al. (2017) comment: 'Complexity is created where risk culture and crisis meet – where risk events (the financial crisis) rearticulate social expectations about the 'end' and exert normative power in re-orienting organizational practices' (p. 155).

The point here is that performance of risk culture is an organizational process and a source of intervention to balance differing prescriptions. In that regard, risk culture will become increasingly important to organizational management of climate risk given the additive potentiality of new risk exposures and the negotiation among actors of risk culture performance.

4.10   Clash of Cultures

Third, observations in the Palermo study were made of the 'blending of existing performance culture with newly produced changes in risk culture.' This interfacing tension of cultures problematized risk practices, according to the authors. Studied research participants faced with new prescriptions to adjust practices to 'clean up their risk culture' underwent a protracted initial phase of organic and idiosyncratic activity of 'connecting the dots' to better understand the implications of prior behavior and the importance of adopting improved risk practices. The problem arose when monetarized incentive-driven actors demonstrated an unwillingness to identify risks to risk culture actors, suggesting the more performance-oriented organizational culture is, the less likely actors will want to criticize risk policies and risk controls which they view as insufficient or are not perceived to be working as well as expected.

This raises another theoretical speculation about culture mix and the resolving of tensions where risk practices involve contributions from all organizational functions and interprofessional practitioners. Viability of mixing performance culture with risk culture pulls focus on a number of questions: What are the implications for rapid risk management efforts under TCFD prescriptions when multiple cultures are concerned? To what extent does the impact of performance-based culture problematize the willingness of actors to talk about their views of corporate limitations, problems, and the open and honest reporting of climate risks? Would performance culture be subsumed into a risk culture, or vice versa?

4.11   Risk Translation

My second tentative proposition of how climate risk is likely to alter informal risk systems involves the practice of risk translation by actors through the translating or 'transforming' of novel risks into more familiar ones. The act of risk translation, in the perspective I am addressing, is done by risk actors certified and trained to identify and assess novel risks so that the organization has a clearer basis and a direction for response action. Previously discussed in Chapters 1 and 2, climate risks are novel risks characterized by uncertainty and unfamiliarity and approximate the idea of a 'manufactured' risk, 'because history provides us with very little previous experience [to understand manufactured risk]' (Giddens, 1999, p. 4).

Moreover, because novel risks are unfamiliar and characterized by uncertainty, they cannot always be readily identified and assessed by traditional risk management approaches. Traditional practices of risk management typically rely on quantitative risk techniques that calculate the probability that risk events will occur and estimate the impact and nature of risk events if they do (Beck, 2006; Giddens, 1999). Uncertain novel and in this case, strategic climate risks, are distinguishable from traditional historical risks which can be calculated by statistical means to calculate frequency and consequence. Approaches, tools, procedures, and structures which work well in the popular and dominant enterprise risk management framework used by many corporations, assist in the assessment and control of relatively simpler historical risks, but are inadequate and may even hamper responsibility dealing with uncertain risks.

Furthermore, and congruent with Power (2009), the possibility exists that climate risks may have the effect of shifting the focus of ERM's focus on rules-based compliance toward an approach addressing the 'critical management of alternative futures' (Power, 2009, p. 852).

4.12   Equivocality of Novel Risks

A further transition challenge is how risk translation responds to the equivocality of novel risks. Take for example, the comprehensive work done by IPCC scientists on global warming, and the remaining view that science does not provide all the answers to when, for example, increasing mean temperatures will precisely occur even though this scenario is key to understanding when and how exactly physical climate risks will materialize. Even in the event of timeline certainty, hydrology experts may still contest the probability and intensity of flooding risk. The exact extent of biodiversity loss may still be disputed by ecologists.

For transition risks, economists may (continue) to argue about the magnitude of climate value at risk and the economic benefits of near term mitigation costs of complex carbon reduction solutions. By extension, and due to lack of quantitative measurement to inform prediction of climate risk events and impacts, precision in cost–benefit analysis of global scale decarbonization costs also becomes problematic.

Characteristics of novel risks imply a level of ambiguity that cannot be easily understood or managed by organizations in the same way historical risks can be. Examples of controllable risk exposures include prior liability risks due to weak risk controls; 'normal' prior hazard risks of slip and fall accidents in facilities, or repetitive company vehicle accidents. Routine, knowable, and established risks are known risks which can be reliably quantified, in contrast to novel 'unknowable,' 'emerging,' or 'incalculable' risks, which cannot (Flage & Aven, 2015; Rudisill, 2013; Van Asselt & Vos, 2008).

Addressing this, Van Asselt and Vos (2008) went as far to say that novel risks cannot be answered by science if '[t]he basic simple questions as to whether it (novel risk) is a real risk, or whether there is enough safety' (p. 282).

Firms responding to novel climate risks seek to understand them in the best way possible but are limited to response actions based on how well they can 'construct' climate risks through a process of risk translation. I proffer this process of risk translation done to reduce equivocality of novel risks may likely be another source of transformation in informal risk practice brought about by the challenge of climate risk. To increase understandings of this, one compelling 20-year case study illustrates how risk translation was constructed by firms responding to the novel risks of the chemical bisphenol BPA (Hardy & Maguire, 2020).

4.13   Impact of Risk Translation

As recorded initially, BPA was not well understood in the business and scientific community and the extent of the risk associated with the use of the chemical in baby bottles, drink thermoses, and storage containers was contested by multiple groups. In 1993 BPA was classified as an endocrine disrupting chemical, potentially adversely affecting the endocrine (hormone) system of animals and humans (Bern et al., 1992; Korach, 1993). The harm was not well understood because the study of endocrine disruption had only begun in the 1990s (Colborn et al., 1996), and current science at the time was divided over how the chemical causes adverse effects.

BPA was, as the authors asserted, a novel unknowable risk, similar in character to an 'emerging risk' (Flage & Aven, 2015). Hardy and Maguire (2020) tracked how initial risk responses among Canadian and Australian scientific communities, regulators, and retailers evolved from an initial state in 1992 of high contestation to increasing acceptance that BPA risks were valid in 2013. Views about the risk of BPA evolved over time through interactions among groups seeking to understand it, culminating in eventual risk translations of BPA as a regulatory, reputational, operational, and strategic risk to those groups exposed to BPA risk. Though unspecified in Hardy and Maguire (2020), the risk translation via the construction of BPA as a 'known' reputational risk would have applied to retailers that later withdrew BPA bottles from store shelves. In another explanatory example, where BPA was constructed as a 'known' strategic risk would have applied to soup can manufacturers which removed BPA linings from products to avoid strategic risk exposure to business growth objectives.

The examples above cast light on a process and the timeline of risk translation that occurred over 20 years culminating in the recognition of BPA as a risk object where the chemical was understood to have potentially caused harm through a verifiable causal pathway and hence became 'understood' as a risk. This is the ongoing, discursive construction process of risk translation, asserted by Hardy and Maguire (2020) containing elements germane to the construction of risk translation, discussed next.

4.14 Construction of Risk Translation

Translating unknown risks into risks that is more readily understood by organizations is a construction process where emerging novel risks are initially constructed as unstable objects of risk knowledge and evolve into a more stable form of risk object. Initial objects of risk knowledge are first contemplated and contested as potentially causing harm and are distinguished from accepted and understood risk objects (Bednarek et al., 2021).

Risk objects are suggested as having three conceptual elements where (1) the object is deemed to pose the risk; (2) the object creates a potential harm; and (3) linkages alleging some form of causation between object and harm are eventually and readily understood. In Hardy and Maguire's (2020) risk translation framework, the elapsed time between novel and risk object status is a liminal state, during which risks still undergo contestations as well as refinements in understandings.

4.15 Translation of Climate Risks

In understanding how climate risks transform informal risk systems, a noteworthy parallel is given between the process of risk translation of BPA with the contemporary risk translation processes for climate risks. In both circumstances, organizational risk actors seek to understand, classify, and translate categories of climate risk into more conventional terms so that risk response actions can be initialized. In the current context, climate risks are approximately classified between risks that are appearing in the near term, i.e., regulatory, reputational, and legal risks versus climate risks that are likely to appear with greater magnitude in the longer time frame, i.e., sea level rising and other extreme weather patterns.

Despite the science supporting the cause of climate change, atmospheric scientists have not yet reached a consensus for the timeline of increasing average temperatures (Knutti & Sedlacek, 2013), suggesting that organizations are operating in a temporary and liminal stage of climate risk translation even though the timeline is key to understanding when and how physical climate risks will materialize. Even if timelines could be agreed upon, it is likely that contestations among specialists over flood risks and biodiversity losses will prevail suggesting that the science actually is not as unequivocal as one might think and that the adverse consequences of physical climate losses are not straightforward.

As Hardy and Maguire (2020) state, 'When organizations face such equivocality, science cannot necessarily lay it to rest. In fact, introducing additional science during this liminal period may simply increase equivocality and contestation' (p. 710). This suggests that the liminal stage is itself at risk of protraction, and possibly delaying the phase at which point risk object formation occurs.

Returning back to the case of PBA, the more that retailers translated equivocality of BPA as a novel risk into reputational risks and take action to lower the carbon footprint of their supply chains, the more their suppliers are likely to translate it as a strategic risk, i.e., one that threatens their business, that might best be by investing in less carbon intensive technologies.

4.16   Risk Roles and Ownership

Another challenge of climate risk in informal risk systems involves interprofessional collaborations of sustainability and financial practitioners in the firm. The term risk management itself is subject to variations in definition but is primarily concerned with the activities of risk practitioners in the identification, assessment, treatment, and monitoring of identifiable risks. However, given the long history of corporate sustainability reporting and the involvement of sustainability professionals to produce such reports, sustainability professionals may be regarded as embedded climate experts.

Yet the standards and stringency of climate risk reporting far exceeds that required in sustainability reporting. Interprofessional challenges may arise between sustainability professionals and financial practitioners in the production of climate risk reporting, given different reporting philosophies. One such contention that is likely to arise is role tensions over constructions of risk materiality. Views of when climate risk is or is not material have direct bearing on disclosure and those views may vary between sustainability and financial practitioners. The prospect of workable collaborations between business impact views held by sustainability practitioners, and the business and financial dependency views held by financial practitioners is another source of transition challenge that climate risk creates in informal systems of risk practice.

4.17 Conclusion

Discussed in this chapter have been three assertions about the transition challenges involving informal risk systems at the firm level. New categories of climate risk will likely bring about transitions in three key areas: (1) adaptations of risk culture to evolve company processes in general; (2) new requirements for risk translation, and (3) new requirements for interprofessional collaborations between sustainability and financial risk practitioners.

References

Beasley, M. S., Clune, R., & Hermanson, D. (2006). The impact of enterprise risk management on the internal audit function. *Journal of Forensic Accounting, 2006*, 1–20.

Beck, U. (2006). *Cosmopolitan vision*. Polity Press.

Bednarek, R., Chalkias, K., & Jarzabkowski, P. (2021). Managing risk as a duality of harm and benefit: A study of organizational risk objects in the global insurance industry. *British Journal of Management, 32*(1), 235–254. https://doi.org/10.1111/1467-8551.12389

Bern, H. A., Blair, P., Brasseur, S., Colborn, T., Cunha, G. R., Davis, W., Dohler, K. D., Fox, G., Fry, M., Gray, E., Green, R., Hines, M., Kubiak, T. J., McLachlan, J., Myers, J. P., Peterson, R. E., Reijnders, P. J. H., Soto, A., Van Der Kraak, G., ... Whitten, P. (1992). Statement from the work session on chemically induced alterations in sexual development: The wildlife/human connection. In T. Colborn & C. Clement (Eds.), *Chemically induced alternations in sexual and function development: The wildlife/human connection* (pp. 1–8). Scientific Publishing.

Bromley, P., & Powell, W. W. (2012). From smoke and mirrors to walking the talk: Decoupling in the contemporary world. *Academy of Management Annals, 6*(1), 483–530. https://doi.org/10.5465/19416520.2012.684462

Colborn, T., Dumanoski, D., & Myers, J. P. (1996). *Our stolen future: Are we threatening our fertility, intelligence and survival?—A scientific detective story*. Dutton Publishing.

Daft, R. L., & Weick, K. E. (1984). Toward a model of organizations as interpretation systems. *Academy of Management Review, 9*(2), 284–295. https://doi.org/10.5465/amr.1984.4277657

Day, R., & Day, J. A. V. (1977). A review of the current state of negotiated order theory: An appreciation and a critique. *Sociological Quarterly, 18*(1), 126–142. https://doi.org/10.1111/j.1533-8525.1977.tb02165.x

Dowbiggin, A. (2018). *Climate risk perceptions in the Ontario (Canada) electricity sector* (Doctoral Dissertation). Heriot Watt University.

Flage, R., & Aven, T. (2015). Emerging risk–Conceptual definition and a relation to black swan type of events. *Reliability Engineering & System Safety, 144*, 61–67. https://doi.org/10.1016/j.ress.2015.07.008

Giddens, A. (1999). Risk and Responsibility. *The Modern Law Review, 62*(1), 1–10.

Gregory, K. L. (1983). Native-view paradigms: Multiple cultures and culture conflicts in organizations. *Administrative Science Quarterly*, 359–376. https://doi.org/10.2307/2392247

Hardy, C., & Maguire, S. (2020). Organizations, risk translation, and the ecology of risks: The discursive construction of a novel risk. *Academy of Management Journal, 63*(3), 685–716. https://doi.org/10.5465/amj.2017.0987

Knutti, R., & Sedláček, J. (2013). Robustness and uncertainties in the new CMIP5 climate model projections. *Nature Climate Change, 3*(4), 369–373. https://doi.org/10.1038/NCLIMATE1716

Korach, K. S. (1993). Surprising places of estrogenic activity. *Endocrinology, 132*(6), 2277–2278. https://doi.org/10.1210/endo.132.6.8504730

Martin, J., & Siehl, C. (1983). Organizational culture and counterculture: An uneasy symbiosis. *Organizational Dynamics, 12*(2), 52–64. https://doi.org/10.1016/0090-2616(83)90033-5

Mikes, A. (2008). Chief risk officers at crunch time: Compliance champions or business partners? *Journal of Risk Management in Financial Institutions, 2*(1), 7–25.

Palermo, T., Power, M., & Ashby, S. (2017). Navigating institutional complexity: The production of risk culture in the financial sector. *Journal of Management Studies, 54*(2), 154–181. https://doi.org/10.1111/joms.12241

Power, M. (2009). The risk management of nothing. *Accounting, Organizations and Society, 34*(6–7), 849–855. https://doi.org/10.1016/j.aos.2009.06.001

Rose, R. A. (1988). Organizations as multiple cultures: A rules theory analysis. *Human Relations, 41*(2), 139–170. https://doi.org/10.1177/001872678804100204

Rudisill, C. (2013). How do we handle new health risks? Risk perception, optimism, and behaviors regarding the H1N1 virus. *Journal of Risk Research, 16*(8), 959–980. https://doi.org/10.1080/13669877.2012.761271

Schein, E. H. (2010). *Organizational culture and leadership*. Wiley.

Silverman, D. (1970). *The theory of organizations: A sociological framework*. Basic Books.

Smircich, L. (1983). Concepts of culture and organizational analysis. *Administrative Science Quarterly, 28*(3), 339–358. https://doi.org/10.2307/2392246

Subramaniam, N., Wahyuni, D., Cooper, B. J., Leung, P., & Wines, G. (2015). Integration of carbon risks and opportunities in enterprise risk management systems: Evidence from Australian firms. *Journal of Cleaner Production, 96,* 407–417.

The Intergovernmental Panel on Climate Change. (2007). *Climate Change 2007: Synthesis Report.* https://www.ipcc.ch/site/assets/uploads/2018/02/ar4_syr_full_report.pdf

The Intergovernmental Panel on Climate Change. (2014). *Climate Change 2014: Synthesis Report.* https://www.ipcc.ch/site/assets/uploads/2018/02/SYR_AR5_FINAL_full.pdf

Van Asselt, M. B., & Vos, E. (2008). Wrestling with uncertain risks: EU regulation of GMOs and the uncertainty paradox. *Journal of Risk Research, 11*(1–2), 281–300.

CHAPTER 5

Formal Risk System Challenges

Abstract Shifts and adjustments to formal risk systems created by climate risk exposure are likely to involve changes in the risk model of the firm, modifications to the practice of risk identification, and assessment to support management decision-making over corporate risk responses. Current risk taxonomies, prioritization criteria, and assessment tools may be rendered inadequate and in need of upgrading to accommodate the new risk category of climate risk.

Keywords Risk model · Risk identification · Risk assessment · Risk indicators · ERM · Risk criteria · Risk taxonomy

5.1 Introduction

In the prior chapter, informal risk systems, understood as the differenti-ated and less tangible systems that exist outside the formal boundaries of risk management practices and structures, were discussed. This included an examination of risk culture, risk translation, and risk ownership chal-lenges at the firm level. Under complexity, and where new risks are not well understood, company actors may involve themselves in acts of risk translation in order to standardize approaches to interpreting risk events and their consequences. Furthermore, contentions among risk actors

© The Author(s), under exclusive license to Springer Nature Switzerland AG 2021
A. Dowbiggin, *Climate Risk and Business*,
https://doi.org/10.1007/978-3-030-78244-3_5

over risk roles and ownership due to collaborative constructions of risk materiality may arise under climate risk circumstances.

Climate risk presents multiple challenges to business through impacts to the formal systems of risk management undertaken by an organization. Of particular emphasis are impacts to the risk model of the firm, process enhancements of risk management with respect to climate risk identification and assessment, and the supporting routines and practices of risk practitioners who bring discipline and commonality to the 'various visions of risk manageability' (Power, 2007, p. 6).

The efforts to formalize risk-oriented 'correlations and sets of principles and administrative procedures' (Scott & Perry, 2012) to support management are ever more in need of examination in light of the call for business to incorporate climate risk into existing risk management systems (TCFD, 2017, 2020a).

Formal risk systems are understood as those systems that are imbued with stable and procedural risk approaches, enabling company actors to rely on preexisting practices of recognizing and assessing risks and in turn, establishing risk strategies in support of overarching corporate strategy and objectives. Formal risk systems are also defined by the use of structured qualitative and quantitative approaches and tools associated with formal risk management frameworks.

However, complexities will likely arise given the non-linear, highly variable, and deep uncertainty of climate risks which are physical and temporally distant, and the climate risks which are transitional and subject to interconnections among market, financial, and legal activities. The integration of climate risk as an entirely new category of risk into formal risk management practices is likely to create transitional changes in a number of ways.

Shifts and adjustments to formal risk systems are likely to involve changes in the risk model of the firm, modifications to the practice of risk identification, and assessment to support management decision-making over corporate risk responses. Current risk taxonomies, prioritization criteria, and assessment tools may be rendered inadequate hence in need of upgrading to accommodate the new risk category of climate risk.

These transitional challenges are discussed in this chapter but preceded first by a preliminary overview of normative risk management with particular emphasis on enterprise-wide risk management, as specified in TCFD guidance documents. The call for companies to incorporate climate risk assessments into 'existing risk management systems' implies that most business practice involves some form of formal risk management systems (TCFD, 2017, 2020a).

5.2 Organization of This Chapter

The organization of this chapter begins with a brief overview of risk management, including enterprise and traditional risk management practices, as approaches to risk identification and assessment differ depending on the framework. Formal structures of risk management are exemplified in the risk model of the firm and reveal linkages and interconnectedness features in organizational life. New transitional challenges relating to new approaches in risk identification and risk assessment are discussed, including how prioritization criteria, risk taxonomies, and systems thinking can assist with climate risk. Opportunities to improve risk management approaches, as well as case examples of climate risk treatments highlighted in risk guidance documents are provided. First, an overview of risk management, discussed next.

5.3 Risk Management

According to risk management guidance documents published by the TCFD in 2020, 'The overarching purpose of risk management is to support a company in achieving its strategy and business objectives. The primary purpose of disclosing risk management processes is to provide context for how the company things about and addresses the most significant risks to successfully executing its objectives and accomplishing its strategy. A company's description of the processes it uses to identify assess and manage risks in general and climate risks in particular may provide readers with some confidence that the company undertakes a rigorous approach to understanding and addressing its risks' (TCFD, 2020b, p. 1).

The call for the integration of climate risks with 'existing risk practices' presupposes firms already rely on a current system of risk management and are reasonably mature in their use of it. Indeed, as TCFD Special Advisor and former U.S. Securities and Exchange Commission (SEC) Chairman Mary Schapiro explained, companies already have a foundation of the relevant skills to make material risk assessments and appropriate disclosures given that companies already have obligations to disclose material risks (Brookbank, 2017).

However, this presumption is potentially problematic. It presupposes a disclosure familiarity and readiness on the part of companies—yet evidence in literature does not categorically suggest this is the case. If climate risk integration is regarded as similar to other business risk integration, one might expect higher levels of corporate climate risk disclosures, aligned with the stringent standards of financial reporting. Disclosures indexed on the CDP suggest a different story during periods prior to and after the TCFD 2017 reporting guidelines were issued.

For example, according to reportage by the Carbon Disclosure Project (CDP), climate risk management practices were underreported, eliciting commentary from Anderson (2019) who noted 'Climate-related risks and other emerging risks are predominantly discussed outside the financial statements' (Anderson, 2019, p. 1). Commenting further on reporting underperformance, O'Dwyer and Unerman (2020) suggest that climate risk system integration is still constrained by 'complexity, allied with limited (or no) control)' (O'Dwyer & Unerman, 2020, p. 14).

Prior to 2017, public entities were forewarned of incoming climate risk disclosure requirements, and some chose to peremptorily report on the TCFD recommendations. However, as Eccles and Krzus (2019) determined in their survey of 15 large oil and gas companies in 2016, only one out of 15 companies reported on each of the 11 recommendations of the TCFD. While addressing all recommendations, the outlier company reported its climate disclosures in a voluntary sustainability report, and not in the recommended financial filing documents (Eccles & Krzus, 2019).

Other empirical work prior to 2017 showed low reporting response for the integration of risk management practice with strategy setting. Corporate survey results suggested company-wide risk management practices accounted for 40% in corporate strategic processes (Ahmad et al., 2014). Those findings followed prior investigatory work done by Beasley et al. (2009), which cautioned ERM implementation was 'highly variable' among firms (Beasley et al., 2005; Beasley et al., 2009).

Given the recent emphasis on climate risk disclosure, these are surprising research findings. The results are even more surprising where the studied groups were large public and private companies with prior sustainability reporting history of climate aspirations, suggesting that their view of climate issues were already of material concern to them (Ahmad et al., 2014; Beasley et al., 2005; Eccles & Krzus, 2019; O'Dwyer & Unerman, 2020).

5.4   Consequences

Recommended climate-related disclosure is intended to be as stringent as 'already in practice' financial disclosures, yet the consequence of disengaged or incomplete reporting will expose business quickly to financial and legal consequences which are already now within view. An example that illustrates financial and legal consequences, is found in the 2021 divestiture decision of the UK's top asset manager Legal & General Investment Management (LGIM).

Announced on June 15, 2021, LGIM reported its divestiture of low carbon disclosing companies, stating in company documents: 'LGIM will divest from AIG, Industrial and Commercial Bank of China (ICBC), China Mengniu Dairy and the US's PPL Corporation for breaching the firm's "red lines" on coal involvement, carbon disclosures or deforestation' (LGIM, 2021). For similar reasons, the divested companies join China Construction Bank, MetLife, Japan Post, KEPCO, ExxonMobil, Rosneft, Sysco, and Loblaw Group, on LGIM's blacklist. This is one example among several high-profile global asset management firms (e.g., others include Blackrock, Inc., Baringa Partners LLP) which have begun announcing their own 'investor-side' risk mitigation of portfolio risk (Blackrock, 2021; Jolly, 2021; Sinclair, 2021).

Prior to the discussion of how climate risk directly impacts formal risk management systems, a brief overview of ERM and its applicability to climate risk management is provided next (TCFD, 2020a).

5.5   ERM Risk Management

Formally stated in the 2020 TCFD Guidance on Risk Management Integration and Disclosure, a preference in writing is made for framing climate-related risks 'using the concepts from a well-recognized and international risk management framework of the Committee of Sponsoring Organizations of the Treadway Commission's (COSO's) enterprise risk management framework' (TCFD, 2020a, p. 3).

What TCFD is referring to is a holistic framework for identifying, defining, quantifying, prioritizing, and treating all material risks of potential loss and gain while simultaneously considering potential correlation and interrelationships between individual risks throughout the organization or enterprise.

ERM was developed by the Committee of Sponsoring Organizations of the Treadway Commission (COSO) in the aftermath of the 2002 US Sarbanes Oxley Act to reform risk management and business reporting by American corporates. ERM represents a leading risk process exemplar which supports organizations in identifying, evaluating, and managing risks at the enterprise level. According to Khan et al. (2016), several factors motivate firms to engage in ERM processes, as follows: the probability of financial distress and associated costs, the low earnings performance, the growth opportunities, and the independence of the board (Blanco-Mesa, 2019; Khan et al., 2016).

In the US, an ERM framework is legally required for financial institutions (e.g., banks, securities brokerage firms, insurance, hedge funds, and mutual funds), but is not legally required for private enterprises. Legally required means required by US statutes, federal case law, or US regulatory agencies (Whitman, 2015). In addition, ERM is an important factor for rating organizations (e.g., in the US, Standard & Poor's), and is accepted by professional organizations as a value-contributing best practice for all enterprises, e.g., in the US, COSO, the Risk and Insurance Management Society (RIMS), and the Casualty Actuarial Society (CAS), which have designed, examined, and promoted ERM frameworks (Whitman, 2015).

Wu and Olson (2009) report the availability of over 80 risk management frameworks globally, with many of them similar to COSO's ERM framework. Despite semantically different definitions, the majority of the 80 frameworks fundamentally provide the same guiding principles of risk identification, assessment, treatment, and monitoring (Wu & Olson, 2009).

5.6   Traditional Risk Management

Another reference in risk management literature, is to 'traditional risk management.' In contrast to ERM, traditional risk management approaches are characterized by siloed and nonintegrated risks most often based on known hazards, and a predisposition to quantification methods based on reliable probability distributions. Unlike ERM, traditional risk management processes tend not to be linked with a strong risk-based mentality among top senior leadership which would otherwise instill risk culture across the firm and embed risk ownership into units, functions, and processes (Lundqvist, 2015).

There is no known literature that considers whether traditional risk management approaches can accommodate climate risks and 'the mapping of them to existing risk categories and types' (TCFD, 2020b, p. 11).

5.7   RISK MODEL OF THE FIRM

An area of transitional challenge to formal risk systems involves the potential refinements of the risk model of the firm. Embedded in the principles of ERM is the notion that companies should have a viable risk model, and under the circumstances of climate risk, appropriate risk models become increasingly more relevant.

Risk models exemplify the formal and structured approach of risk control within a firm. Risk models delineate risk management accountabilities, guide the practices of risk actors, and create an integrated view of how risks are interconnected and managed throughout the enterprise. Most are hierarchical in form, designating risk governance practices as a control level, the office of the CEO as another control level, and risk committees accountable to the CEO as another control level. Other risk models such as the operational three lines of defense (3LoD) model assign risk accountability to business units and are commonly found in the banking and insurance industry (Potter & Toburen, 2016).

As an excerpted example from risk guidance documents of the TCFD, the Italian multinational oil and gas company Eni S.p.A headquartered in Rome with operations in 66 countries, with a market capitalization of US$36.08 billion, explained their corporate risk model this way:

> The company has adopted a new company mission based on the UN SDGs and a definition of strategic guidelines and targets for the energy transition in the short medium and long term. The company has structured governance on climate with a central role of the board in managing main issues connected with climate change, the presence of specific committees, the establishment of the (Climate) Advisory Board and programs focused on climate change issues. The company includes targets related to the energy transition in management incentives, consistent with the medium and long term. (Eni, 2020, p. 46)

5.8   COMPLEXITY OF RISK MANAGEMENT

The point is that climate risk increases the complexity of risk management through additive and compounding effects. Effective management of them requires internal control and coordination that a well-designed formal risk model can foster (Elliot, 2018). Risk roles are defined in the model as are expectations about which groups are responsible for the identification and assessment of climate risks. Coordination among levels of risk control in the model can be augmented by internal control and risk management systems (ICRMS) facilitating decision rules, procedures, and organizational structures aimed at identifying, measuring, and later managing risks that are climate related. Furthermore, a risk model of the firm engenders the principle of 'interconnectedness' articulated in climate risk guidance discussed next.

5.9   INTERCONNECTEDNESS

As already introduced, one of the related challenges to business is likely to be in the 'integration of climate risk identification and assessment into existing risk practices' (TCFD, 2020b). Successful integration is fostered by transitional risk system changes that promote interconnectedness among company actors that may differ from current practices. As TCFD states, '[t]he principle of interconnections means all relevant functions, departments and experts are involved in the integration of climate risks into the company's risk management processes and in the ongoing management of climate related risks' (TCFD, 2020b, p. 18).

While it may be up to academics and consultants to identify proxy measures of interconnectedness for performance measurement, reasonable intuition suggests that involving all company actors in the pursuit of climate risk recognition and management will necessarily require cognitive cohesiveness fostered by interconnectedness. Operationalizing the interconnectedness construct will require even greater effort on the part of businesses which have relied on older, more traditional risk management approaches. Shifting away from siloed views of risk management and ownership by business unit or department for example will likely represent the greatest trajectory of transition while ERM organizations, given their orientation to cross company coordination, may have an easier transition, relatively speaking.

Furthering the principle of interconnectedness in an enterprise view of climate risk will necessarily involve other transitions. In formal risk management systems, the first stepwise methodological approach is to begin with risk identification, discussed next.

5.10 CLIMATE RISK IDENTIFICATION

In practice, processes of how risk actors agree to define the qualitative features of climate risks and their impacts on strategy are open to revision and represent distinct shifts in the process steps of risk management.

Related to climate risk identification are fundamental ideas about how to categorize knowable risks according to their qualitative features. Identification practices include the consideration of the events or trends that create the risk exposure, including the risk driver i.e., extreme climate events, carbon pricing regimes, or unevenness in governmental climate policy pathways.

5.11 APPROACHES TO RISK IDENTIFICATION

Risk identification differs in practice between traditional risk practices and ERM. All risks may be identified by the positional source of risk drivers and contemplated as either endogenous or exogenous risks. If endogenous risk events occur inside firm boundaries, then they are thought to be controllable through organizational control mechanisms. If risk drivers are externally located, hence exogenous, then intraorganizational risk control mechanisms are more difficult to implement. Traditional risk management is largely oriented toward the identification of risks occurring within organizational boundaries. The evolution of risk management practices to address the additional risks occurring in external business environments have had the effect of containing traditional practices to 'insurable' risks where downside risks can be formally and financially transferred. The impacts related to the risk drivers associated with climate risk are in the majority outside the control of companies, and as such, designated as strategic risks (Elliot, 2018).

5.12 NEW APPROACHES TO IDENTIFICATION

In the ERM approach, recognition is given to both risks and opportunities occurring in both internal and external business environments. Given the focus on downside implications in this book, 'climate risk opportunities' are outside the scope of discussion.

As previously discussed, the complex and systemic nature of climate risk to affect multiple systems and entities simultaneously or in a cascading manner will necessitate new approaches to risk identification (Sidortsov, 2014). For example, transition risks of declining market interest in company products or weakening supply chain relationships due to extreme weather impact on logistics performance will have knock-on effects producing amplified risk exposures related to corporate reputation, stakeholder perceptions, and potential credit and legal liability risks (Carney, 2021).

Another scenario illustrates how multiple risks identification is complicated by uncoordinated risk strategies of the firm. For example, the use of carbon reduction technologies in product innovation (an opportunity) is negated by poor emissions reporting potentially triggering legal liability exposure. Such risk scenarios illustrate the complexity and systemic nature of climate risk and its identification, in contrast to past traditional and earlier ERM approaches to view recognized risks as 'standalone' exposures.

5.13   Climate Risk Taxonomy

In further discussion about risk identification, schemes that can be logically applied to an agreed upon risk taxonomy of climate risk will likely become critical. While a risk taxonomy is the typically hierarchical categorization of risk types done in ERM processes, a common approach is to adopt a tree structure, whereby risks higher in the hierarchy are decomposed into more specific manifestations (Elliot, 2018). Constructing a risk taxonomy follows the practice and science of general taxonomies, classifying things or concepts, including the principles that underlie such classifications. The outcome is a classification scheme, or a formal list of risk types (Carr et al., 1993).

Identification measures conducted across the enterprise through various ERM techniques, e.g., workshops, Delphi technique, surveys, and scenario analysis, may produce variations in identified risk categories and subcategories problematizing taxonomy or classification schemes for climate risk, as well as other business risks the company is exposed to (Elliot, 2018).

This may occur due to the propensity of company actors to prioritize risks closest to them operationally, under 3LoD models for example. Risks may also be prioritized according to expected time of onset. A problem arises when the greatest exposures lie in the far future and when emphasis on near term exposures is prioritized, contributing to what Carney (2021) calls 'the tragedy of the horizon.'

Company exposures to climate risk are variable to time of onset and clear identification of that factor is needed in risk identification schemes. What makes it even more problematic is that all climate risks are strategic in nature and uncontrollable by organizational measure, suggesting that effective prioritization of risks is of paramount importance.

5.14   CLIMATE RISK AS RISK FACTOR

Creating a taxonomy of climate risks in the risk identification stage is further complicated by views appearing in literature of climate risk as being not just a risk, but as a risk factor that amplifies existing business risks. Risk factors in practice are viewed as early warning signals to anticipate risk before it happens. Risk factors are furthermore correlational with existing risk but regarded as not necessarily causal, in the sense that 'correlation does not prove causation.'

At a global scale, the examination of risk factors has been applied to various 'high-stakes' scenarios. Research studying the linkages between physical climate risks and armed conflict has been addressed for example, in the work of Mach et al. (2019). Their research asserts that the intensification of climate change is estimated to increase future conflict risk, and that already "3 to 20 per cent of conflict risk over the last century has been influenced by climate variability or change" (Mach et al., 2019, p. 3).

Climate risk-commerce linkages in supply chain management intuitively suggest a similar picture, though empirical evidence in literature is still nascent with some exceptions. Tenngren et al.'s (2020) research on climate risk within global logistics management, referred to in practice as 'Transnational Climate Impacts,' considers the secondary economic disruptions worldwide from hurricane systems and extreme flooding (Haraguchi & Lall, 2015; Tenggren et al., 2020). "The knock-on effects of global interconnectedness exceed(ed) the primary damage from extreme weather events and caused a ripple effect through the world trade system" (Tenggren et al., 2020, p. 1266). Given potential economic losses, Tenggren et al.'s 2020 study of 14 Swedish large exporting companies nevertheless found low proactive risk management among the participant group going as far as to suggest that some were operating under the misapprehension that risk was being outsourced to Tier 1 suppliers by contingency stock requests, despite suppliers not having the ability nor capacity to mitigate for climate risks.

This example illustrates that continued extreme weather disruptions to the operations of global suppliers would imperil local sourcing and transportation routes and create volatility in supplier management. The point is, companies already manage 'normal' risks within supply chains, e.g., procurement, pricing, and delivery performance—but in the additive overlay of weather-related climate risks, risk exposures of business interruption intensify under such circumstances.

5.15   Systems Thinking

One risk thinking tactic in support of viewing climate risk as an intensifying risk factor is to adopt a system thinking approach. Systems thinking is a skill and an approach that would help risk actors and decision-makers acknowledge that the effect of climate risk cuts across other risks (O'Donnell, 2005).

Systems thinking is a holistic big picture perspective that, as an analogy, focuses on the forest instead of the trees (Richmond, 2000). However, building on Senge (1990), a more appropriate objective for evaluating risks is to see both the trees and the forest. A systems-view of risks make it possible to see through the complexities and to illuminate the interdependency of recognized climate risks.

5.16   Temporality of Risks

Another complication of climate risk in the identification phase relates to the temporality and speed of onset of climate risk types. Many ERM systems are aligned with business planning cycles normally three to five years in duration, and not with the longer timelines understood with climate change. This vexatious issue, articulated by (Carney, 2020, 2021) as the 'tragedy of the horizon,' highlights recency bias and the problem of insufficient present-day climate risk mitigation to protect future economic and natural systems.

Climate risks that are physical in nature, as documented in the IPCC's Fifth Assessment Report (AR5) and in Mach et al.'s (2016) analysis of AR5, have an almost certain trajectory of increasing temperatures, sea level rises, floods, and other deleterious consequences of accumulating atmospheric GhG concentrations. Mach et al. (2016) assert that 'the potential for risk reduction (mitigation) will change over the next few decades, an era with some further committed warming, and in the second half of the twenty-first century and beyond, a longer-term era of climate options determined by the ambition of global mitigation' (Mach et al., 2016, p. 427).

Squaring assertions made by Mach et al. (2016) with Carney (2021) reveal estimates of risk severity have changed dramatically in reportage in the 5 years between report dates. This may signify the errant and deeply uncertain nature of climate risk. For business, this has direct implications in formal risk management practice with respect to estimations of the time of onset of risks, let alone their systemic consequences for business operations.

How risks are recognized qualitatively and classified by risk type is an integral part of the ERM risk management process. Identifying risks that can be classified as climate risks will exist in risk taxonomies alongside of other business risks. Applying ERM approaches to identify and classify climate risks is a critical first step to the assessment of climate risks, discussed next.

5.17   Climate Risk Assessment

A company's assessment of its risks is fundamental to its prioritization of those risks and management of the most significant ones. The work of practitioners assessing climate risks according to conventional ERM approaches may encounter newly found challenges to account for how transition risks—the more immediately manifested risks relative to physical risks—are assessed.

5.18   Risk Prioritization

One problem that may arise is in the prioritization of risks, and in the use of criteria against which risks are prioritized. Normative practitioner literature emphasizes a traditional likelihood and impact matrix to assess the severity of risks. Risks are identified as determined by the juncture of their likelihood and impacts and then evaluated on the basis of severity relative to the firm's risk appetite to tolerate those risks.

Given the temporality of physical climate risks and the likelihood of risks to manifest in different timelines, the addition of 'speed of onset criteria,' i.e., the elapsed time between the risk event occurring and the point at which the company first feels the effect of it could better inform company efforts in developing risk response plans.

5.19   OTHER RISK FACTORS

Additionally, other risk factors that intensify risk effects, such as: (1) vulnerability; and the (2) exposure of the business; (3) the operational hazard; and (4) the interacting stressors that are antagonistic or additive to risk events occurring, should be considered. An explanation of each follows.

The 'vulnerability' criteria to gauge the level of susceptibility of a company to a risk event can better prepare the organization to reduce the impact—if the event occurs. Vulnerability is closely associated with impact and likelihood, but the addition of these criteria (business exposure, operational hazard and interacting stressors) can help the firm prepare and adapt relative to the level of vulnerability the organization has (TCFD, 2020b). In practice, risk exposure is a quantified loss potential of business. Traditionally, risk exposure is usually calculated by multiplying the probability of an incident occurring by its potential losses. This may serve businesses well to understand what financial exposure of the business is at stake given the onset of a climate risk event. In the case of a climate risk that is 'transitional' i.e., declining market interest in company carbon-based products, the business may be relatively less exposed if it readily has low carbon product substitutes being brought to market.

Operational hazard can be viewed as an important risk factor in assessment and is usually defined as a source of harm in working conditions. Under warming temperatures, poor working conditions due to insufficient air conditioning would be a risk factor intensifying physical risks brought to bear on human capital and property degradation (Hayes, 2017).

Interacting stressors, a term more routinely used in ecological study, relates to additional factors existing in the environment which amplify risks. In the natural world, chemicals e.g., pesticides, metals, and excess nutrients, such as nitrogen released by industry and farming that are added to soil environments, are viewed as chemical stressors that interact with the environment (U.S. EPA, 2019). Climate risk stressors may add to the risk or complicate the management of risks, depending on its origin.

Formal risk management systems can be improved with additional prioritization criteria, as the above example illustrates. In ERM practices, risks are often represented in risk heat maps, graphically showing the relative location of risks. A certain risk may have both low impact and low likelihood on a traditional heat map, but when reappraised with additional criteria, emerges as a severe risk needing prioritization. For example, the speed of onset and vulnerability are two additional criteria, but organizations could develop other criteria relevant to their sector or firm characteristics.

5.20   Scenario Planning

Furthermore, climate risks which are conceptualized as occurring in the far future, defy formal methods of risk planning—which may assume future conditions will be an extended trajectory of current conditions. Scenario planning, conducted in the risk identification and assessment phases, model a number of future plausible worlds, where business dependencies on climate effects are considered and reported (Cobb & Thompson, 2012). Explicitly noted in TCFD documents is the call for companies to report on their scenario plan in four possible future world eventualities (Haigh, 2019). Scenario planning to the scale and detail needed for climate risk disclosure is not currently a common business practice. Corporate capacity is challenged to meet the needs of scenario planning which requires expert skills and a level of expertise needed in firms assessing their risk exposures to a range of possible futures. Indeed, as the Climate Disclosures Standards Board in a survey found, only one out of 80 companies disclosed the results of their scenario analysis (Topping, 2012).

5.21   Conclusion

This chapter discusses the relevance of a risk model for an organization and the organization of risk processes under traditional and enterprise-wide risk management frameworks. Process transitions in formal risk management systems relating to climate risk identification and assessment are presented. With climate risk, the process steps of the first two phases in normative risk management practice would necessarily involve changes: firstly, how climate risk is identified and qualitatively distinguished from other business risks, and secondly, how climate risks are prioritized as strategic risks and supported by evaluations which include the use of key risk factors (KRF). KRFs, if designed appropriately, can assist in the indication of probable onset of climate risk events, speed of onset, organizational impact location, and in the improvement of the organization's risk profile. The potential value of KRFs to strategic risk management of climate risks that are physical and transitional appears conceptually under-reported in literature yet represents a significant source of transition in climate risk practice.

Strategic risks in practitioner literature are partially qualified by risk sources, especially from external factors outside the organization that would interfere with the company's strategic objectives. Risk control of strategic risks is extremely limited for businesses hence the value and necessity of KRFs as early warning signals of emerging strategic risks. In other literature, however, strategic risks are defined by both location and 'potentiality' specified by Frigo and Anderson (2011) who assert: 'Strategic Risk Management is a process for identifying, assessing and managing risks and uncertainties, affected by internal and external events or scenarios, that could inhibit an organization's ability to achieve its strategy and strategic objectives with the ultimate goal of creating and protecting shareholder and stakeholder value' (Frigo & Anderson, 2011, p. 22).

In practice, exemplified by case examples used in TCFD risk management guidance documents, companies designate climate risk as a strategic risk though approaches to managing them appear to vary (TCFD, 2020b).

Understanding climate change risk requires seeing this full picture. The key risk assessment enabled consideration of complex interactions, where some characteristics and processes shaping hazards, vulnerability, and exposure can be measured quantitatively and modeled probabilistically, while others cannot.

References

Ahmad, S., Ng, C., & McManus, L. A. (2014). Enterprise risk management (ERM) implementation: Some empirical evidence from large Australian companies. *Procedia-Social and Behavioral Sciences, 164*, 541–547. https://doi.org/10.1016/j.sbspro.2014.11.144

Anderson, N. (2019). *IFRS Standards and climate-related disclosures*. International Financial Reporting Standards. https://www.pwc.ch/en/publications/2021/in-brief-climate-change-nick-anderson.pdf

Beasley, M. S., Branson, B. C., & Hancock, B. V. (2009). ERM: Opportunities for Improvement Take your risk management system to the next level. *Journal of Accountancy, 208*(3), 28.

Beasley, M. S., Clune, R., & Hermanson, D. R. (2005). Enterprise risk management: An empirical analysis of factors associated with the extent of implementation. *Journal of Accounting and Public Policy, 24*(6), 521–531. https://doi.org/10.1016/j.jaccpubpol.2005.10.001

Blackrock. (2021). *Net zero: A fiduciary approach.* https://www.blackrock.com/corporate/investor-relations/blackrock-client-letter

Blanco-Mesa, F., Rivera-Rubiano, J., Patino-Hernández, X., & Martínez-Montana, M. (2019). The importance of enterprise risk management in large companies in Colombia. *Technological and Economic Development of Economy, 25*(4), 600–633. https://doi.org/10.3846/tede.2019.9380

Brookbank, D. (2017, June 29). *TCFD Analysis: The focus now turns to re-thinking 'materiality'.* Responsible Investor. https://www.responsible-investor.com/articles/tcfd-analysis-materiality

Carney, M. (2020). *TCFD: Strengthening the Foundations of Sustainable Finance.* https://www.suerf.org/policynotes/9319/tcfd-strengthening-the-foundations-of-sustainable-finance

Carney, M. (2021). *Value(s): Building a better world for all.* Penguin Random House Canada.

Carr, M. J., Konda, S. L., Monarch, I., Ulrich, F. C., & Walker, C. F. (1993). *Taxonomy-based risk identification.* Software Engineering Institute, Carnegie-Mellon University.

Cobb, A. N., & Thompson, J. L. (2012). Climate change scenario planning: A model for the integration of science and management in environmental decision-making. *Environmental Modelling & Software, 38,* 296–305.

Eccles, R. G., & Krzus, M. P. (2019). Implementing the task force on climate-related financial disclosures recommendations: An assessment of corporate readiness. *Schmalenbach Business Review, 71*(2), 287–293. https://doi.org/10.1007/s41464-018-0060-4

Elliot, M. (Ed.). (2018). *Risk Management Principles and Practices 3rd Edition ARM 54.* The Institutes.

Eni. (2020). *Annual Report 2020.* https://www.eni.com/assets/documents/eng/reports/2020/Annual-Report-2020.pdf

Frigo, M. L., & Anderson, R. J. (2011). What is strategic risk management? *Strategic Finance, 92*(10), 21.

Haigh, N. (2019). *Scenario planning for climate change: A guide for strategists.* Routledge.

Haraguchi, M., & Lall, U. (2015). Flood risks and impacts: A case study of Thailand's floods in 2011 and research questions for supply chain decision making.

International Journal of Disaster Risk Reduction, 14, 256–272. https://doi.org/10.1016/j.ijdrr.2014.09.005

Hayes, J. (2017). *Operational decision-making in high-hazard organizations: Drawing a line in the sand.* CRC Press.

Jolly, J. (2021, January 26). BlackRock threatens to shed some of its massive fossil fuels portfolio. *Mother Jones.* https://www.motherjones.com/environment/2021/01/blackrock-threatens-to-shed-some-of-its-massive-fossil-fuels-portfolio/

Khan, M. J., Hussain, D., & Mehmood, W. (2016). Why do firms adopt enterprise risk management (ERM)? *Empirical Evidence from France. Management Decision, 54*(8), 1886–1907. https://doi.org/10.1108/MD-09-2015-0400

Legal and General Investment Management. (2021, June 15). *LGIM renews pressure on companies to provide climate accountability and achieve net-zero emissions.* https://www.legalandgeneralgroup.com/media-centre/press-releases/lgim-renews-pressure-on-companies-to-provide-climate-accountability-and-achieve-net-zero-emissions/

Lundqvist, S. A. (2015). Why firms implement risk governance–Stepping beyond traditional risk management to enterprise risk management. *Journal of Accounting and Public Policy, 34*(5), 441–466. https://doi.org/10.1016/j.jaccpubpol.2015.05.002

Mach, K. J., Kraan, C. M., Adger, W. N., Buhaug, H., Burke, M., Fearon, J. D., Field, C. B., Hendrix, C. S., Maystadt, J., O'Loughlin, J., Roessler, J. S., Schultz, K. A., & von Uexkull, N. (2019). Climate as a risk factor for armed conflict. *Nature, 571*(7764), 193–197. https://doi.org/10.1038/s41586-019-1300-6

Mach, K. J., Mastrandrea, M. D., Bilir, T. E., & Field, C. B. (2016). Understanding and responding to danger from climate change: The role of key risks in the IPCC AR5. *Climatic Change, 136*(3–4), 427–444. https://doi.org/10.1007/s10584-016-1645-x

Material Handling & Logistics. (2017, September 4). *Supply chain impact of Hurricane Harvey could be worse than expected.* https://www.mhlnews.com/global-supply-chain/article/22054536/supply-chain-impact-of-hurricane-harvey-could-be-worse-than-expected

O'Donnell, E. (2005). Enterprise risk management: A systems-thinking framework for the event identification phase. *International Journal of Accounting Information Systems, 6*(3), 177–195. https://doi.org/10.1016/j.accinf.2005.05.002

O'Dwyer, B., & Unerman, J. (2020). Shifting the focus of sustainability accounting from impacts to risks and dependencies: Researching the transformative potential of TCFD reporting. *Accounting, Auditing & Accountability Journal, 33*(5), 1113–1141. https://doi.org/10.1108/AAAJ-02-2020-4445

Potter, P., & Toburen, M. (2016). The 3 lines of defense for risk management. *Risk Management, 63*(5), 16.

Power, M. (2007). *Organized uncertainty: Designing a world of risk management.* Oxford University Press.

Richmond, B. (2000). *The "thinking" in systems thinking.* Pegasus Communications.

Scott, S., & Perry, N. (2012). The enactment of risk categories: The role of information systems in organizing and re-organizing risk management practices in the energy industry. *Information Systems Frontiers, 14*(2), 125–141.

Senge, P. (1990). *The fifth discipline: The art and practice of the learning organization.* Doubleday.

Sidortsov, R. (2014). Reinventing rules for environmental risk governance in the energy sector. *Energy Research & Social Science, 1,* 171–182. https://doi.org/10.1016/j.erss.2014.03.013

Sinclair, D. (2021, June 21). *The Journey to Net Zero: Why this is more tan just EMR 2.0.* https://www.baringa.com/en/insights-news/points-of-view/the-journey-to-net-zero-why-this-is-more-than-just/

TCFD. (2017). *Recommendations of the Task Force on Climate-related Financial Disclosures.* https://assets.bbhub.io/company/sites/60/2020/10/FINAL-2017-TCFD-Report-11052018.pdf

TCFD. (2020a). *2020 Status Report.* https://www.fsb.org/wp-content/uploads/P291020-1.pdf

TCFD. (2020b). *Guidance on Risk Management Integration and Disclosure.* https://www.fsb.org/wp-content/uploads/P291020-2.pdf

Tenggren, S., Olsson, O., Vulturius, G., Carlsen, H., & Benzie, M. (2020). Climate risk in a globalized world: Empirical findings from supply chains in the Swedish manufacturing sector. *Journal of Environmental Planning and Management, 63*(7), 1266–1282. https://doi.org/10.1080/09640568.2019.1660626

Topping, N. (2012). How does sustainability disclosure drive behavior change? *Journal of Applied Corporate Finance, 24*(2), 45–48.

U.S. EPA. (2019). *Climate Change and Interacting Stressors: Implications for Coral Reef Management in American Samoa* (Final Report). U.S. Environmental Protection Agency, Washington, DC, EPA/600/R-07/069.

Whitman, A. F. (2015). Is ERM legally required? Yes for financial and governmental institutions, no for private enterprises. *Risk Management and Insurance Review, 18*(2), 161–197. https://doi.org/10.1111/rmir.12045

Wu, D. D., & Olson, D. L. (2009). Introduction to the special section on "optimizing risk management: Methods and tools." *Human and Ecological Risk Assessment, 15*(2), 220–226. https://doi.org/10.1080/10807030927 60967

Response, Materiality, and Disclosure Challenges

Abstract The organization of this chapter addresses risk response actions, materiality, and reporting and disclosure challenges. Formal risk management process challenges exist in the risk treatment of climate risks even when guided by prioritization criteria to establish 'the top list of strategic climate risks' facing an organization. Complications also arise from new materiality considerations, climate risk reporting, and disclosure challenges, discussed in this sixth and final chapter.

Keywords Risk Response · Materiality · Climate Disclosure · TCFD · Climate reporting · Risk Model · Risk management process · Emissions reporting · Carbon management · Carbon accounting

6.1 Introduction

The prior chapter discusses the relevance of an organizational risk model and the arrangements of risk processes under traditional and enterprise-wide risk management frameworks. Process transitions in formal risk management systems relating to climate risk identification and assessment are presented. With climate risk, the process steps of the first phases in normative risk management practices involve changes firstly, how climate risk is identified and qualitatively distinguished from other

© The Author(s), under exclusive license to Springer Nature
Switzerland AG 2021
A. Dowbiggin, *Climate Risk and Business*,
https://doi.org/10.1007/978-3-030-78244-3_6

business risks, and secondly, how climate risks are prioritized as strategic risks and supported by evaluations, including the use of key risk factors (KRF). KRFs, if designed appropriately, can assist in timing indication of probable onset of climate risk events, speed of onset, and organizational impact location. As a consequence, KRFs assist in improvement of the organization's risk profile. The potential value of KRFs to the strategic management of climate risks that are physical and transitional appears conceptually underreported in literature yet represents a significant opportunity for process transition in climate risk practice.

Also discussed were strategic risks in practitioner literature that are partially qualified by risk source—that being from external factors outside the organization and that would interfere with the company's strategic objectives. Risk control surrounding strategic risks is extremely limited for business hence the necessity of KRFs as early warning signals of climate-related threats. Risk scholars in other literature define KRFs by potentiality in addition to risk source. For example, Frigo and Anderson (2011) assert: 'Strategic Risk Management is a process for identifying, assessing and managing risks and uncertainties, affected by internal and external events or scenarios, that could inhibit an organization's ability to achieve its strategy and strategic objectives with the ultimate goal of creating and protecting shareholder and stakeholder value' (p. 84). Case examples used in TCFD risk management guidance documents reflect preferences of companies to designate climate risk as strategic though reported approaches to managing them appear to be variable (TCFD, 2020a).

6.2   Organization of This Chapter

The organization of this chapter follows ERM framework approaches to address the third risk management phase: risk response, or how the company will treat risk(s). Process challenges exist in the risk treatment of climate risks even when guided by prioritization criteria to establish 'the top list of strategic climate risks' facing an organization. Complications arising from new materiality considerations and emergent climate risk reporting and disclosure practices are discussed in this sixth and final chapter.

6.3   Risk Response Action

Risks are identified and assessed and then placed in a risk taxonomy to illustrate and communicate the qualitative distinctions of recognized risks. Risks are further prioritized in a stepwise fashion according to prioritization criteria germane to the organization and its risk appetite. When sufficient decision-useful risk data is collected and expert judgment applied, organizations enter into the actionable phase of decision making over approaches to mitigate risks. Given the characterization and classification of climate risks as strategic, climate risk is a risk category resistant to most internal risk control efforts. Companies cannot control extreme weather events nor the transition risks of policy and regulatory changes or financial and credit risks created by future debt and creditor renunciation. The alternative response is to mitigate the risks through various response approaches manifested as the risk strategy of the firm. Response actions result from internally generated decisions about how the entity prefers to respond to risk; however, ontological standardization of risk created by institutional sources dictates that business will in fact mitigate against climate risks. This is a compulsory response requirement regardless of corporate preference.

While the mitigative risk response choice is now an emergent standard for business in many and increasing numbers of jurisdictions, variations in the strategic objectives of how the company will enact mitigation options are left up to the firm.

The scope, if any, of mitigation options, i.e., the climate risk strategic objectives therefore becomes an important element of corporate climate mitigation strategy. The potential scope of mitigative response options raises questions about the choices a company makes, the resulting effects of those choices on carbon performance, and the more salient consideration of whether the company's mitigative efforts actually result in desired outcomes.

6.4   Scope of Response

In literature, prior research indicates variation in corporate climate strategies pertaining to decarbonization efforts. Decarbonization efforts are exemplified new process transitions relating to (1) 'unlocking' carbon in intra organizational systems through energy and materials transition; (2) tracking and reducing carbon through emissions reporting, carbon management, and carbon accounting practices; and (3) preventing further use of carbon through lifecycle assessment processes.

Response action challenges which incorporate a broad solution set of strategic risk responses imply that companies may or may not know the potential effect of effort, i.e., how much is enough. This may be due to time lags in effects and the absence of carbon budgeting against which sectoral and individual firm effort can be measured.

6.5   TIME HORIZON

A further risk response challenge is in the prolongation of strategic efforts beyond conventional ERM planning cycles of two to five years. Planning horizons for risk response aligned with ERM methods need now to be pushed out to time horizons where effects of severe physical climate risks are expected to materialize. Adjustment to accommodate longer time horizons is a source of process challenge in the risk response phase of climate risk management.

Materiality considerations underlying other challenges facing business with respect to the reporting and disclosure practices of climate risks are discussed next. Materiality is a key component of how organizations evaluate and prioritize the risks it faces, as well as determining whether those risks should be disclosed.

6.6   MATERIALITY CHALLENGES

Materiality of the resulting risks identified in ERM practices is critical to the way the company uses risk analysis and incorporates it into risk management processes. Materiality is also important in both the undertaking and reporting on climate risks regardless of whether they can be quantified or not. As Anderson (2019) suggests, 'Qualitative external factors, such as the industry in which the company operates, or investor expectations may make some risks "material" warranting disclosures in financial statements, regardless of their numerical impact' (p. 3).

In corporate reporting traditions, materiality has been formally defined through practice as the ascertainment of the financial importance of the disclosure as an item of information to primary users, i.e., shareholders, investors, and creditors. The stricture of materiality is codified by professional financial practitioner organizations and upheld by investor audiences which need to understand the risk profile of investable assets.

In financial reporting domains, risks which can impact the financial profile of an entity now or in the near term must be reported in compliance with institutional rules and reporting regulations (Directive, 2014/56; Regulation, 537/2014; Sarbanes-Oxley Act, 2002).

In nonfinancial reporting, however, materiality considerations have largely been guided by a raft of sustainability reporting groups as 'issues' which companies can elect to comment on to a greater or a lesser extent. Accountability and obligation to exercise judgment on the materiality of nonfinancial issues has fostered corporate sustainability reporting as a source of nonbinding, standalone and in many jurisdictions, unverified claims (O'Dwyer & Unerman, 2020). Materiality approaches used in prior empirical work also demonstrated low levels of systematic assessments of climate change related issues and environmental reporting (Anderson, 2019). While some groups e.g., Sustainability Accounting Standards Board (SASB) attempt to standardize materiality of nonfinancial issues for companies and sectors, problems remain with materiality judgments relating to climate risk disclosure.

Principally, five challenges arise, as discussed next.

6.7   FIVE MATERIALITY CHALLENGES

Firstly, a lack of guidance on how management should apply judgment to which climate issues are material has been problematized in literature (O'Dwyer & Unerman, 2020). A special project of the CDP, the not-for-profit Climate Disclosure Standards Board (CDSB), indicates that climate issues should be treated as material if they give rise to financial impacts in the short term, or over longer times scales or if they threaten the resiliency of corporate strategies or business model or affect its ability to generate or preserve value. However, the TCFD defines materiality of climate risk as an 'all-in' proposition, given that all climate-related risk is non-diversifiable (TCFD, 2017). This is likely to become contentious at the firm level, if materiality assessments of certain climate-related risks (evaluated by internal corporate actors) render them immaterial, but yet from a reporting point of view, nevertheless need disclosure (O'Dwyer & Unerman, 2020).

A second concern noted in financial literature is the apparent failure of reporting companies to explain how materiality was established. Existing materiality approaches were raised as a concern by the 2020 report on risk management integration and disclosure by the TCFD which notes that 'companies frequently failed to explain the process by which they determine the materiality of climate related risks to their operations' (TCFD, 2020a).

A third contention involves the relative time frame in which risk exposures become material. Under climate risk disclosure guidelines, there is an implied requirement for the use of process-oriented approaches to materiality judgment, to determine issues which may not currently be material but have the potential to become 'materially-relevant' in the future. This calls for forward thinking and a view of materiality likelihood that may defy current corporate reporting skills and practices. The materiality of future nonfinancial issues can only be assessed by companies on a qualitative basis, rendering the long-term view of them highly speculative.

A fourth challenge occurs when issues actually become material and in the absence of guidance on it, creating additional complications. To illustrate this point, the 2019 TCFD status report found that 60% of 198 preparers viewed climate risks as material now or in the next one to two years while only 19% considered them to be material in the next three to 10 years (TCFD, 2019). This suggests that materiality judgments can vary among corporations, and that climate risk reporting may be driven by corporations' needs to reflect their own judgments over those of external audiences' expectations (O'Dwyer & Unerman, 2020).

A fifth challenge to be contemplated is the creation process of materiality judgment where inter-profession interactions between sustainability professionals and financial corporate reporters are concerned. Insight into how materiality is constructed and enacted in complex business situations has been studied by Power (1996) and Pentland (1993) who deemed that materiality is constructed in enigmatic environments where different reporting philosophies are enacted. Differences in reporting outlook e.g., business impacts vs corporate dependencies on climate, and different targeted audiences e.g., financial investors vs broader stakeholders are likely to interfere with materiality construction. Requirements for interprofessional collaboration and productive exchanges between sustainability and financial reporters will become necessary.

The above five challenges of materiality judgment reveal a picture that suggests current nonfinancial materiality practices have limitations to accommodate climate risk reporting. While all corporate risk reporting practices have limitations, the practice of climate risk materiality judgment is still evolving. At the time of writing, however, it remains to be seen how materiality judgments will be standardized.

The next point of focus is on the inevitable reporting transitions that are likely to occur, once materiality judgments on which 'risks to report' have been made.

6.8   Reporting Transitions

The practice of disclosing business risk is not new to corporate reporters; however, new challenges emerge for climate risk reporting, as indicated in financial and accounting literatures. Two issues revealed in literature underscore problematic areas of disclosure reporting. These areas relate to transitional shifts and adjustments in disclosure reporting practices, as a consequence of new climate risk reporting requirements (TCFD, 2017, 2018, 2019, 2020b). One is the philosophical shift from an impact orientation to climate dependency orientation in reporting. Second is the attendant skill and expertise needed to evaluate and report on business dependencies on climate as part of overall scenario-based modelling conducted in the risk assessment phase.

Disclosure practices, theory, and the challenges arising for companies with long histories of sustainability reporting in their transition to climate risk reporting are discussed next.

6.9   Climate Risk Disclosure

Risk disclosure, or 'the making of risk information public' in alignment with reporting frames is a specialized business activity involving contributions from risk actors and investor communication specialists, to make known the approaches the company is taking with respect to 'a) processes for identifying and assessing climate-related risks...b) processes for managing climate related risks...c) how above processes are integrated into overall risk management' (TCFD, 2017, p. 14).

Companies marshal skills and expertise to construct formal risk reporting for investor and stakeholder audiences, once formal risk evaluations of recognized climate risks are conducted, and management decision making over appropriate risk responses are complete. Climate risk disclosure involves company reporting of the multiple exposures to physical and transition risks. Carbon disclosures, in particular, involve communicating to third-party audiences the committed actions of companies to measure and manage the reduction of emissions through carbon management programs. Risk governance disclosures impart information on how climate risk oversight is achieved while the financial disclosures of the economic effect of climate on company operations are disclosed through scenario analysis and other methods. This is not an exhaustive list of the subtypes of disclosures management may elect to disclose but it nonetheless provides a brief tentative view of corporate disclosure. Why companies disclose is discussed next.

6.10   Disclosure Theory

Explanatory theories about why companies disclose are broadly divided among three theoretical groups: (1) social-political theories; (2) economic theories of voluntary disclosure; and (3) institutional theories (Hahn et al., 2015).

In the context of the first, social-political theory, companies are deemed to have response requirements to external social or political pressures exerted by stakeholders or the desire to maintain corporate legitimacy in the judgmental view of external audiences (Hahn & Lulfs, 2014).

The second theoretical group is economics-based theories of disclosure, where companies signal to interested groups, information based on evaluations of costs and benefits (Clarkson et al., 2008). According to signaling theory arguments, 'reporting potentially reduces the principal-agent problem of asymmetric information by increasing transparency' (Connelly et al., 2011).

The third theoretical anchor to the topic of corporate disclosure is institutional theory, where organizations are driven not merely by their aim to maximize profits, but also by the requirements of different institutions (e.g., governments, institutional investors).

Institutional forces manifested by the TCFD climate risk reporting guideline in combination with institutional stakeholder pressure and government carbon policies establish a reasonably clear case that institutional theory anchors why public companies are or will disclose. Economic-based or signaling explanations do not directly apply given the central principle driving TCFD's reporting guidelines of 'the availability and provision of high quality understandable and reliable information' (Carney, 2015, p. 3; Unerman et al., 2018) Social-political theory of social pressure and legitimacy drives nonfinancial sustainability reporting but lacks full explanatory power for climate risk disclosure.

6.11   Climate Risk Related Disclosure

Transitional challenges associated with emergent disclosure guidelines disclosures are appreciated when contrasted with sustainability emissions reporting practices over the past 20 years. Corporate reporting is not new. Today and likely tomorrow's climate risk disclosure practices are both similar and different from sustainability and emissions reporting, discussed next.

6.12   Sustainability Reporting

Prior to the release of 2017 TCFD guidelines, reporting companies responded voluntarily to nonbinding disclosure guidelines promulgated by numerous sustainability reporting groups (e.g., Global Reporting Initiative (GRI), Ceres, Climate Disclosures Standards Board (CDSB), Carbon Disclosure Project (CDP), Sustainability Accounting Standards Board (SASB), United Nations Sustainable Development Goals (UNSDG), United Nations Principles for Responsible Investment (UNPRI), and the European Union Guidelines on reporting climate-related information). The stated objective of this reporting was generally to improve corporate transparency by producing an integrated picture of nonfinancial reporting alongside financial disclosures.

Sustainability reporting communicates information about nonfinancial issues and impacts, in Corporate Social Responsibility (CSR), environmental, social and governance (ESG) frames in a decidedly impact reporting orientation. Companies select one or several frameworks to communicate information to relevant audiences to promote green agendas or encourage capital flows to low carbon initiatives. Sustainability reporting encourages an accounting for organizational intention, strategy, and expected performance related to nonfinancial measures, by following a framework or standards provided by over 25 reporting groups (Eco Act, 2021).

For example, GRI calls for sustainability reporting according to its proprietary set of standards pertaining to how ESG issues are impacted by business activities. Other reporting groups (e.g., CERES, CDSB, CDP, SASB) offer alternate standards addressing ESG, some with variable weightings of business impact. The plethora of sustainability frames ultimately created an 'alphabet soup' of reporting frameworks (Antoncic, 2019) leading to frustration and criticism by corporate groups

(IFRS Foundation, 2019). Contention in academic literature furthermore suggests quality standards are low, few, voluntary, and non-standardized, and fails to provide a 'broad range of investors or lenders with information about the potentially substantial risks to financial returns resulting from a corporation's dependencies on climate' (O'Dwyer & Unerman, 2020, p. 2).

Corporations and investors have commented on the plethora of nonfinancial reporting frameworks guidelines and standards causing confusion and disclosure overload and raised the risk of climate risk disclosure being 'swamped with a large volume of potentially contradictory and confusing impact-oriented disclosures' (World Business Council for Sustainable Development, 2019, p. 4). Summing up these frustrations, Hans Hoogervorst, the International Accounting Standards Board's chair (IFRS Foundation, 2019) said: '...there are simply too many standards and initiatives in the space of sustainability reporting. This leads to a lot of confusion among users and corporations themselves. To give one example, Tesla is ranked highest in terms of the sustainability index of MSCI, while FTSE ranks it as the worst carmaker globally on ESG issues. Yet another agency puts it somewhere in the middle. People may be forgiven for not making heads or tails of it. Moreover, with so many standards, the potential for disclosure overload is enormous' (International Financial Reporting Standards, 2019).

6.13   Emissions Reporting

Emissions reporting, including the audit and verification of GhG emissions of companies, was adopted with the onset of emissions trading schemes (ETS) in some countries. Reliable assurances were and are needed when carbon crediting and emission exchanges have transactional value between entities.

In literature, the introduction of ETS schemes in Europe and the US provoked studies of carbon accounting and related disclosures (Bowen & Wittneben, 2011; Dawkins & Fraas, 2011; Haigh & Shapiro, 2012; Milne & Grubnic, 2011; Prado-Lorenzo et al., 2009; Ziegler et al., 2011).

While the longer term contribution of ETS activities of carbon crediting and emissions exchanges is in serious doubt (Colmer et al., 2020), measurement and verification skills acquired to satisfy ETS schemes may usefully transfer to TCFD emissions reporting requirements. Specifications of how GhGs are to be reported in climate risk disclosures are different and designated in scope zones, formulated by the Greenhouse Gas Protocol. GhG emissions in Scope 1 zones cover the direct emissions from owned or controlled sources; Scope 2 covers indirect emissions from the generation of purchased electricity, steam, heating, and cooling consumed by the reporting company and Scope 3 includes all other indirect emissions that occur in a company's value chain (Carbon Trust, 2021).

6.14   Disclosure Skills Transfer?

Climate risk reporting recommendations set out more explicit and narrower guidelines for disclosing climate-related risks to the extent that companies must account for the financial impacts on company business created by warming scenarios. In contrast to other frameworks, TCFD reporting guidelines produce an expectation of reporting performance to the level of stringency consistent with compulsory financial reporting. Skills and expertise acquired by firms for prior sustainability reporting schemes and for prior emissions reporting into trading schemes, while not financial, may nonetheless provide a preliminary basis for climate disclosure.

6.15   Conclusion

A critical factor for the credibility of climate risk management is mandatory disclosure under TCFD guidelines. At the time of writing, mandated reporting is to be enacted in the UK, and expected shortly in other sovereign jurisdictions including the European Union, Singapore, Canada, Japan, and South Africa, and New Zealand. Additionally, sub-national groups with jurisdiction over regional capital and insurance markets (e.g., Province of Ontario, Canada, The New York State Department of Financial Services (DFS)), as well as other state, local governments, not for profit and public sector groups, have formally announced expectations that corporate financial filings must be in alignment with TCFD guidelines.

Future prospects of advanced disclosure practices involving due diligence, third-party assurances, and the specter of regulatory consequence are yet to be realized. While these distinct prospects are looming, the full extent of climate risk challenges addressed in this book is consequential enough for business.

References

Anderson, N. (2019). *IFRS Standards and climate-related disclosures*. International Financial Reporting Standards. https://www.pwc.ch/en/publications/2021/in-brief-climate-change-nick-anderson.pdf

Antoncic, M. (2019). Why sustainability? Because risk evolves and risk management should too. *Journal of Risk Management in Financial Institutions, 12*(3), 206–216.

Bowen, F., & Wittneben, B. (2011). Carbon accounting: Negotiating accuracy, consistency, and certainty across organizational fields. *Accounting, Auditing & Accountability Journal, 24*(8), 1022–1036.

Carbon Trust. (2021). *Briefing: What are Scope 3 emissions?* https://www.carbontrust.com/resources/briefing-what-are-scope-3-emissions

Carney, M. (2015, September 29). *Breaking the tragedy of the horizon-climate change and financial stability.*

Clarkson, P. M., Li, Y., Richardson, G. D., & Vasvari, F. P. (2008). Revisiting the relation between environmental performance and environmental disclosure: An empirical analysis. *Accounting, Organizations, and Society, 33*(4–5), 303–327. https://doi.org/10.1016/j.aos.2007.05.003

Colmer, J., Martin, R., Muûls, M., & Wagner, U. J. (2020). *Does Pricing Carbon Mitigate Climate Change? Firm-level Evidence from the European Union Emissions Trading Scheme.* Centre for Economic Performance, London School of Economics and Political Science.

Connelly, B. L., Certo, S. T., Ireland, R. D., & Reutzel, C. R. (2011). Signaling theory: A review and assessment. *Journal of Management, 37*(1), 39–67. https://doi.org/10.1177/0149206310388419

Dawkins, C., & Fraas, J. W. (2011). Coming clean: The impact of environmental performance and visibility on corporate climate change disclosure. *Journal of Business Ethics, 100*(2), 303–322. https://doi.org/10.1007/s10551-010-0681-0

Directive 2014/56. Amending Directive 2006/43/EC on statutory audits of annual accounts and consolidated accounts. European Parliament, Council of the European Union.

Eco Act. (2021). *The Big eBook of Sustainability Reporting Frameworks.* https://info.eco-act.com/hubfs/0%20-%20Downloads/Sustainability%20Reporting%20Frameworks%20-%20eBook/EN/The%20Big%20eBook%20of%20Sustainability%20Reporting%20Frameworks%20-%20EN.pdf

Frigo, M. L., & Anderson, R. J. (2011). Strategic risk management: A foundation for improving enterprise risk management and governance. *Journal of Corporate Accounting & Finance, 22*(3), 81–88. https://doi.org/10.1002/jcaf.20677

Hahn, R., & Lülfs, R. (2014). Legitimizing negative aspects in GRI-oriented sustainability reporting: A qualitative analysis of corporate disclosure strategies. *Journal of Business Ethics, 123*, 401–420.

Hahn, R., Reimsbach, D., & Schiemann, F. (2015). Organizations, climate change, and transparency: Reviewing the literature on carbon disclosure. *Organization & Environment, 28*(1), 80–102. https://doi.org/10.1177/1086026615575542

Haigh, M., & Shapiro, M. A. (2012). Carbon reporting: Does it matter? *Accounting, Auditing & Accountability Journal, 25*(1), 105–125.

International Financial Reporting Standards. (2019). *Speech: IASB Chair on what sustainability reporting can and cannot achieve.* https://www.ifrs.org/news-and-events/2019/04/speechiasb-chair-on-sustainability-reporting/

Milne, M. J., & Grubnic, S. (2011). Climate change accounting research: Keeping it interesting and different. *Accounting, Auditing & Accountability Journal, 24*(8), 948–977.

O'Dwyer, B., & Unerman, J. (2020). Shifting the focus of sustainability accounting from impacts to risks and dependencies: Researching the transformative potential of TCFD reporting. *Accounting, Auditing & Accountability Journal, 33*(5), 1113–1141. https://doi.org/10.1108/AAAJ-02-2020-4445

Pentland, B. T. (1993). Getting comfortable with the numbers: Auditing and the micro-production of macro-order. *Accounting, Organizations and Society, 18*(7–8), 605–620. https://doi.org/10.1016/0361-3682(93)90045-8

Power, M. (1996). Making things auditable. *Accounting, Organizations and Society, 21*(2–3), 289–315. https://doi.org/10.1016/0361-3682(95)00004-6

Prado-Lorenzo, J. M., Gallego-Alvarez, I., & Garcia-Sanchez, I. M. (2009). Stakeholder engagement and corporate social responsibility reporting: The ownership structure effect. *Corporate Social Responsibility and Environmental Management, 16*(2), 94–107. https://doi.org/10.1002/csr.189

Regulation 537/2014. On specific requirements regarding statutory audit of public-interest entities and repealing Commission Decisions 2005/909/EC. European Parliament, Council of the European Union. https://eur-lex.europa.eu/legal-content/EN/TXT/?uri=celex%3A32014R0537

Sarbanes-Oxley Act of 2002, Pub. L. No. 107–204, 116 Stat. 745 (2002). https://www.congress.gov/bill/107th-congress/house-bill/3763

TCFD. (2017). *Recommendations of the Task Force on Climate-related Financial Disclosures.* https://assets.bbhub.io/company/sites/60/2020/10/FINAL-2017-TCFD-Report-11052018.pdf

TCFD. (2018). *2018 Status Report.* https://assets.bbhub.io/company/sites/60/2020/10/FINAL-2018-TCFD-Status-Report-092518.pdf

TCFD. (2019). *2019 Status Report.* https://assets.bbhub.io/company/sites/60/2020/10/2019-TCFD-Status-Report-FINAL-0531191.pdf

TCFD. (2020a). *Guidance on Risk Management Integration and Disclosure*. https://assets.bbhub.io/company/sites/60/2020/09/2020-TCFD_G uidance-Risk-Management-Integration-and-Disclosure.pdf

TCFD. (2020b). *2020 Status Report*. https://assets.bbhub.io/company/sites/60/2020/09/2020-TCFD_Status-Report.pdf

Unerman, J., Bebbington, J., & O'Dwyer, B. (2018). Corporate reporting and accounting for externalities. *Accounting and Business Research, 48*(5), 497–522. https://doi.org/10.1080/00014788.2018.1470155

World Business Council for Sustainable Development. (2019). *Reporting matters—Navigating the landscape: A path forward for sustainability reporting*. https://www.iasplus.com/en-ca/news/regulations/2019/report ing-matters-navigating-the-landscape-a-path-forward-for-sustainability-rep orting

Ziegler, A., Busch, T., & Hoffmann, V. H. (2011). Disclosed corporate responses to climate change and stock performance: An international empirical analysis. *Energy Economics, 33*(6), 1283–1294. https://doi.org/10.1016/j.eneco.2011.03.007

General Conclusion of the Book

The full extent of climate risk challenges to business, as envisioned in this work, has been presented as a fivefold proposition. How firms expect to initiate and accelerate their own organizationally bounded mitigative approaches to reduce climate risk and reach net zero targets will likely be varied, vast and complex. In the thought experiment of this work, climate risk will likely prompt transitions in organizational cognitive and risk belief systems, organizational resources, informal and formal management systems, and the way in which firms conceptualize and disclose future business dependencies on climate. Each of these tentative propositions is summarized in the general conclusion of this book, addressed next.

Cognitive and Risk Belief Challenges

The first contemplation addressed in this work is the cognitive transition organizations will most likely undergo. As discussed, the emergence of climate risk appearing on multiple fronts in business implies a need for different ways of regarding risk in organizations, resulting in what theorists call a 'cognitive transformation in a complex setting.' As shown, climate risk complicates business in many ways, starting with how decision makers think or construe future risks beyond their personal experience and current knowledge of their company's business exposure and vulnerability. The attainment of higher order complexity thinking is likely to

© The Author(s), under exclusive license to Springer Nature
Switzerland AG 2021
A. Dowbiggin, *Climate Risk and Business*,
https://doi.org/10.1007/978-3-030-78244-3

emerge, as a result of the need for elevated management cognition and risk beliefs by company actors.

Why should this matter? Lack of attention to a particular risk exposure, or some aspect of the risk, is often driven by the belief that all exposures have been identified. Low cognitive processing of risk, or reluctance to perceive the importance of prioritizing one risk over another, may limit and possibly amplify unattended risks, leading to negative outcomes. An example in theory given was the case of not recognizing the sequencing of the onset of all systemic risk events: relatively less harmful risks occur first and are quickly controlled, delaying management attention to more harmful risks occurring in the near future. In practice one might contemplate the case of procurement practice reform, through carbon emissions monitoring as an example, as a means to manage climate risk in the supply chain, all the while physical climate risks have exposed some suppliers creating the probability of supply interruptions for the firm.

Also addressed was the cognitive impact of climate risk on organizations where cognitions and belief systems enable firm level action and sustain organizational change processes. This proposition is heavily supported in the general management cognitions literature, as shown, though the study of cognitive transformation of company actors when facing the full spectra of climate risk has not been addressed, to the best of my knowledge.

Many opportunities exist in research to understand the full implications of climate risk on organizational cognition and risk belief given that extant organizational theories have not considered these implications either in theory or empirical work. Business and management theory could be enlivened on this issue through the pursuit of certain research questions or approaches, such as those below.

Research questions—Cognition and risk belief challenges

1. Which approaches in phenomenological research could best examine the lived experience of company actors to account for how they think about future systemic climate risks?
2. Are the cognitions of company actors of future-bound events consistent across business groups or sectors?
3. What might account for the differences, if any?
4. Are there comparative differences in cognition and risk belief about climate risk within the firm among different professionals

i.e., risk managers and financial managers versus sustainability managers?

5. What organizational strategies and techniques do companies utilize, if any, to enhance homogeneity of climate risk perceptions to advance shared purpose?

These and other research questions that focus on the holistic portfolio of climate risks (as opposed to just 'climate change'), can assist both academic and practitioner groups in understanding how organizations can manage the unfolding crisis of climate risk impacts on business.

RESOURCE CHALLENGES

As addressed in Chapter 3, carbon reduction programs require unprecedented resources to identify, collect data, measure, analyze, and report on carbon emissions. This may include contingent resources to evaluate and acquire other complimentary assets, processes, and dynamic capabilities needed in support of decarbonization efforts. Task loads for existing corporate actors or the human capital costs of new expertise in engineering, financial, and specialist consulting, will likely exceed the status quo. This is probable even for corporations previously reporting on emissions data for pollution regulators. Now, under new and more anticipated climate risk disclosure requirements for large reporting companies, and for the smaller corporations embedded in carbon chains of larger ones, new and different resources are required. However, the range of current resource-based theories—understood as those theories which focus on the internal factors with organizations and emphasize the firm's own decisions and competencies is paltry. Understanding how organizations will determine, assess, and utilize resources in support of corporate climate risk mitigation objectives is not well explored in literature. The more enduring resource theory and its variants attributed to Hart's (1995) original theory has yet to catch up with climate risk. Applying resource-based theory to the domain of climate risk is notably absent in literature, at the time of writing. Climate risk exposures for companies is a new application domain for resource theories, and as such, it challenges prior explanations. It also prompts one to contemplate a number of research questions, provided below.

Research questions—Resource challenges

1. Given that the RBV and NRBV theories are based on impacts-out views of business activity and hence incongruent with the double materiality of climate risk (see Chapter 1), can RBV theory accommodate climate risk?
2. In what manner can resource-based theories be refurbished to account for decarbonization strategy, and what empirical approach would sufficiently interrogate this?
3. Can resource based theories provide explanatory power for corporate duality and paradox (where organizations grapple with risk reduction AND risk taking for corporate business growth at the same time)?
4. Given low levels of prior work on the novelty of climate risk as an application domain for resource theory, various exploratory research designs will likely be most fitting. Case study work or ethnographic approaches could 'pull focus' on these resource challenges for corporations.

INFORMAL RISK SYSTEM CHALLENGES

The challenge of climate risk for business has also been shown to create transitional pressures on informal risk systems within business organizations. In Chapter 4, I proposed that risk management needs to be distinguished as being both an informal and a formal risk system. Perspectives on this distinction were inspired from literatures for risk practitioners, as well as accounting, and management literature. Practical insights into the problems of resolving contention in risk culture shifts, the ambiguity of novel risks and the arising tensions from inter professional involvement in climate risk analysis and reporting were discussed in some detail.

I posited these three issues as occurring in the informal risk system, understood as the differentiated and less tangible systems that exist outside the formal boundaries of risk management practices.

The introduction of new climate risk categories to a corporation will leave an indelible mark on its risk culture. Companies once observed as having unbridled approaches to business growth now face constraints and obligations to comply with stringent climate risk regulations. How businesses reconcile their internal risk cultures with new requirements is not well understood. An example of risk culture transformation was provided using interpretative research done by researchers of financial groups after the UK banking crisis. I proposed that the 'messy life cycle' of risk culture reform could be applied theoretically to the decarbonizing corporation. However, and overall, the study of risk culture and of risk culture reform brought about by increasing institutional complexity and far-reaching risk regimes, i.e., TCFD, is thinly understood and almost absent in climate literature. This lack of research into risk culture transition raises questions about the process and scope of transition companies may experience with climate risk.

Research questions—risk culture transitions

1. How can risk culture reform and transition be observed? What sequential changes take place in risk culture reform, and which methodology best suits the empirical setting, and why?
2. Which, if any, organizational drivers or antecedents were instrumental in other studied cases, and can they be applied to the probable reforms in risk culture of companies mitigating against climate risk?
3. What are the critical success factors for risk culture change to accommodate climate risk?

Another element I proposed in the informal risk system of the organization pertains to risk translation by actors to 'understand' climate risk. Relevant research on organizational translation of ambiguous and not well understood novel risks into more concretized risk 'object' states was offered by Hardy and Maguire (2020). Their 20-year longitudinal work studied networked organizations which responded to the novel risk of the industrial chemical bisphenol A (BPA). I proffered that the same process of risk translation is another source of transformation in informal risk practice brought about by the challenge of climate risk. It is a noteworthy parallel given between the process of risk translation of BPA with the contemporary risk translation processes for climate risks. In both circumstances, organizational risk actors seek to understand, classify, and translate categories of climate risk into more conventional terms so that risk responses can be initialized and communicated. It also provokes ideas about how climate risk translation by organizations might unfold. Several research questions addressing this are included below.

Research questions—risk translation

1. How do actors translate climate risk management approaches into requirements for corporate strategy?
2. How do organizations translate climate risk into reputational risk or legal liability risk? How do organizations translate elements of climate risk from novel risk states into risk objects?
3. How are translation processes constructed?

The third element discussed in Chapter 3 related to the challenge of climate risk in informal risk systems involving interprofessional collaborations of sustainability and financial practitioners in the firm. By way of explanation, the term risk management itself is subject to variations in definition but is primarily concerned with the activities of risk practitioners in the identification, assessment, treatment, and monitoring of identifiable risks. However, given the long history of corporate sustainability reporting and the involvement of sustainability professionals to produce such reports, sustainability professionals may continue to be regarded as climate risk experts. However, the standards and stringency of climate risk measurement, assessment and reporting exceeds those required in prior corporate sustainability reporting, raising questions about the need for specialized expertise to conduct climate risk evaluations and reporting activities.

As a consequence, inter professional challenges may arise between sustainability professionals and financial practitioners in the production of climate risk assessment and reporting, given different reporting philosophies. One such contention likely to arise are role tensions over constructions of risk materiality. Views of when climate risk is or is not material has direct bearing on disclosure and those views may vary between sustainability and financial practitioners. The prospect of workable collaborations between business 'impact' views held by sustainability practitioners and the business and financial 'dependency' views held by financial practitioners is another source of transition challenge, and one that could be further illuminated in academic research. Questions addressing this may emerge as best practices guidance for practitioners to explicate the following:

Research questions—risk roles and ownership

1. What are the critical success factors for successful climate risk management, as far as practitioner qualifications are concerned?
2. What constitutes a climate risk expert team within the organization?
3. How do differences in climate risk reporting philosophies get reconciled?
4. Should climate risk assessors also be climate risk reporters for the organization?

FORMAL RISK SYSTEM CHALLENGES

As discussed in Chapter 5, climate risk presents multiple challenges to business through impacts to the formal systems of risk management undertaken by an organization. Of particular emphasis in this chapter were three concerns: a) impacts to the risk model of the firm; b) process enhancements of risk management with respect to climate risk identification and assessment; and c) the supporting routines and practices of risk practitioners who bring discipline and commonality to formal risk practice.

Why should this matter? As discussed, formalizing risk-oriented practices that support management is evermore in need of examination in light of the call for businesses to incorporate climate risk into existing risk management systems. In other words, the way in which companies identify manage and integrate climate risk into corporate strategy is a key requirement under TCFD reporting. Noncompliance with this mandate would most likely result in other risks for the corporation, including reputation, legal, and investor risks.

This chapter discussed the relevance of a risk model for an organization and the organization of risk processes under traditional and enterprise-wide risk management frameworks. Process transitions in formal risk management systems relating to climate risk identification and assessment were presented. With climate risk, the process steps of the first two phases in normative enterprise risk management practice would necessarily involve certain modifications. They are firstly, to establish how climate risk is identified and qualitatively distinguished from other business risks, and secondly, how climate risks are prioritized as strategic risks and supported by evaluations which include the use of key risk factors (KRF). As discussed, KRFs, if designed appropriately, can assist in the indication of probable onset of climate risk events, speed of onset, organizational impact location, and in the improvement of the organization's risk profile. The potential value of KRFs to strategic risk management of climate risks that are physical and transitional were noted as being conceptually underreported in literature yet represented a significant source of transition in climate risk practice.

Furthermore, strategic risks in practitioner literature were noted as partially qualified by risk source, that being from external factors outside the organization, that would interfere with the company's strategic objectives. Risk control surrounding strategic risks is extremely limited for businesses hence the value and necessity of KRFs as early warning signals of near term and emergent strategic risks. In other literature, however, strategic risks were defined by both location and potentiality.

Many of those issues addressed in Chapter 5 invite questions relating to the systemic nature of climate risk and its management by organizations.

Response, Materiality, and Disclosure Challenges

In the final chapter, the ERM framework was again used to frame the discussion about the third risk management phase: risk response, or how the company will treat climate risk(s). Challenges were noted regarding the processes followed in the risk treatment of climate risks even when guided by prioritization criteria to establish 'the top list of strategic climate risks' facing an organization. Complications were also noted as arising from new materiality considerations and emergent climate risk reporting and disclosure practices. Of further note was the challenge organizations face in the prolongation of strategic efforts beyond conventional ERM planning cycles of two to five years. Congruently with other researchers, I proposed that planning horizons for risk response aligned with ERM methods need to be pushed out to time horizons where effects of severe physical climate risks are expected to materialize. Adjustments to accommodate longer time horizons is a source of process challenge in the risk response phase of climate risk management. Efforts to mitigate recency bias by extending time horizons are done to avoid what former Bank of England Chairman Mark Carney called the 'tragedy of the horizon.' To understand the complexity of climate risk response, materiality considerations and reporting and disclosure practices more meaningfully, several research questions (by no means an exhaustive list) are suggested below.

Research questions—response, materiality, and disclosure challenges

1. What modifications to enterprise risk management methodology are needed to accommodate strategic climate risks? Is the stepwise framework of ERM still relevant, and how might formal risk practices change? (It should be noted that risk monitoring is the 4th 'leg' of ERM methodology.)
2. From an empirical standpoint, what approaches do actively decarbonizing corporations take to monitor their efforts and to determine whether their mitigative efforts actually result in desired outcomes?
3. From an empirical standpoint, given carbon reduction targets far exceed the length of time managers are likely to remain working, what mechanisms can the firm use to ensure continuity of mitigative effort?
4. What is the extent of sectoral guidance on climate risk mitigation for individual companies, and what impact might that have on competitive advantage among firms in the same sector?

Summary

While there is much empirical work to be done and many other research questions to be asked, I view them as necessary in a critically needed exploratory line of enquiry. By raising questions that ask how business is truly coping with new risk categories of climate risk and in what ways their organizational practices and priorities have shifted, we gain a deeper and richer understanding of the future of climate mitigation by business.

INDEX